Handbook of Sustainable Refurbishment

The Handbook of Sustainable Refurbishment

Housing

Simon Burton

publishing for a sustainable future

London • New York

First published 2012
by Earthscan
2 Park Square, Milton Park, Abingdon, Oxon OX14 4RN

Simultaneously published in the USA and Canada
by Earthscan
711 Third Avenue, New York, NY 10017

Earthscan is an imprint of the Taylor & Francis Group, an informa business

British Library Cataloguing in Publication Data
A catalogue record for this book is available from the British Library

Library of Congress Cataloging in Publication Data
Burton, Simon, 1945-
  Handbook of sustainable refurbishment : housing / Simon Burton. — 1st ed.
      p. cm.
  Includes bibliographical references and index.
1. Energy consumption. 2. Dwellings—Remodeling. I. Title.
  HD9502.A2B88 2011
  333.79'6316—dc22
                                        2011008843

ISBN 978-1-84971-198-2 (hbk)

Typeset in Bembo
by FiSH Books, Enfield, UK

Contents

List of Figures and Tables

Tables

Acknowledgements

Thanks to the following individuals and organizations who, among many others, have specifically contributed to this book: Charlie Baker (Urbed); Deutsche Energie-Agentur (DENA); John Doggart (Sustainable Energy Academy); Energy Intelligent Education; Energy Savings Trust; David Gottfried (US Green Building Council); Sarah Harrison (Eco-Refurbishment); Homes and Communities Agency; Ulla Janson; Paul Jennings (ALDAS); Pierre Levy (Detry & Levy); Lucy Pedler (Archipeleco Architects); Andy Simmonds (Simmonds Mills Architects); and Russell Smith (Parity Projects); Jes Mainwaring, and Robert Cohen.

Thanks are also extended to the following architects: Ulf Brannies and Rita Fredewess (Oldenburg); Michael Felkner (Hopferau); Martin Gruber (Eichstetten); Paul and Johannes Hettlich (Cologne); Andreas Hipp (Leipzig); Andreas Koch (Constance); Bernd Melcher (Mannheim); and Gebr. Schmidt (Ahrensburg).

I am equally indebted to all those providing photographs, particularly: Markus Bachmann; Eicken and Mack; Thomas Dix; Jeffery Pike; and James Brittain.

Preface

The starting point for this book is the acceptance that man-made climate change is happening, with potentially disastrous consequences, and that the reduction of carbon dioxide emissions from the burning of fossil fuels is a world priority. The buildings sector is a major energy user and, as a part of this, housing providers, owners and users must take urgent steps to reduce energy use wherever possible.

There are a large number of existing homes in most countries that waste energy due to old or poor construction and inefficient operation. New dwellings are constructed with much higher energy standards due to regulations, new products and a general awareness that building energy use constitutes a large proportion of national use and must be reduced. While new housing will make an increasing contribution to reducing carbon emissions from energy use, the existing stock cannot be replaced in any reasonable timescale, even if this is seen as desirable. The debate over demolition versus refurbishment in different situations includes issues of cost, conservation of heritage, environmental impacts, community stability, life-cycle analysis, etc., and can result in demolition and new buildings as the best solution. However, this can never eliminate the need to renovate vast amounts of the existing housing stock in order to reduce its carbon emissions.

This book provides some answers to how this can be achieved through organizational and technical methods. Nevertheless, this is not a problem that can be solved overnight. It is worth stating why existing dwellings are generally not renovated to desirable sustainability standards. There are four main reasons: money, ignorance, disruption and lack of legislation.

Money is spent on housing refurbishment (in the UK alone this amounts to UK£4 billion a year); but not much is allotted to actions that reduce carbon emissions from energy use – most goes to essential repairs, redecoration and new bathrooms and kitchens. Money spent on energy-use reduction often does not have a visible return, and the payback on reduced fuel bills usually takes many years to transpire. Finance may also not be available, or may be difficult and expensive to raise via grants or loans. Ignorance on the part of occupants, builders and craftsmen on how to reduce energy use is widespread. Building work creates disruption to established routines, hassle to life, in general, and dirt and dust. While new construction is becoming highly regulated in terms of its sustainability (and this also applies to some aspects of replacement and refurbishment in some countries), the regulation of existing housing stock to make it sustainable is a minefield.

Worldwide, the housing market is a complex web of location, ownership and tenure, age, type, construction, state of repair, and use, so that a range of actions is necessary to create sustainable housing at a national level. Energy is used in housing for heating, cooling, lighting, cooking, appliances and ventilation; all can be reduced and related methods are addressed in this book.

The easiest components of the jigsaw puzzle are the technical solutions needed to create sustainable housing; these have been well researched and information is easily available from government-supported organizations and system manufacturers, and in professional and DIY design guides and manuals. It is the organizational aspects of raising awareness, strategies, management and finance that provide the major stumbling blocks at present.

This book is aimed at those who work in local authorities, housing departments and social housing organizations, as well as at architects and building consultants. It is based on research carried out over numerous years and draws on good-practice case studies, outlining the technical fixes and background information, methodologies, financing, etc., which lead to successful conclusions.

No apology is made for recommending high insulation standards and choosing as far as possible case studies where very low energy consumption has been achieved. The case studies include *PassivHaus* examples (the original German system for very low-energy houses) that have been achieved in both single-family housing and blocks of flats. While this is not always feasible in all situations, the purpose is to point the way ahead – to show that real sustainable refurbishment is possible and is already happening. The scale of the problem of energy use in existing housing is such that minor increases in loft insulation and a few low-energy bulbs are not enough: radical improvements are essential.

Looking ahead, many governments have the vision of very low- or zero-carbon housing within the next 50 years. This will require a vast effort by everybody involved, both in the individual houses themselves and within each country, if all existing housing stock is to be successfully treated.

List of Acronyms and Abbreviations

ACH	air changes per hour
ADEME	Agence de l'Environnement et de la Maîtrise de l'Energie
AECB	The Sustainable Building Association
BFRC	British Fenestration Rating Council
BREEAM	BRE Environmental Assessment Method
CERT	Carbon Emissions Reduction Target
CHP	combined heat and power
CO_2	carbon dioxide
DHW	domestic hot water
DIY	do it yourself
DPC	damp-proof course
e	emissivity
EPBD	Energy Performance of Buildings Directive
ERDF	European Regional Development Fund
ESCO	energy service company
EU	European Union
FSC	Forest Stewardship Council

GBC	Green Building Council
GLA	Greater London Authority
HID	high-intensity discharge
HQE	*Haute Qualité Environnementale*
INCA	Insulated Render and Cladding Association
LED	light-emitting diode
LEED	Leadership in Energy and Environmental Design
microCHP	micro-combined heat and power
MVHR	mechanical ventilation system with heat recovery
OSB	oriented strand board
Pa	pascal
PAYS	Pay As You Save scheme
PHPP	*PassivHaus* Planning Package
PU	polyurethane
PV	photovoltaic
TRV	thermostatic radiator valve
UK	United Kingdom
US	United States
VAT	value-added tax
WC	water closet

1 Outline of Sustainable Housing Refurbishment

The refurbishment of housing takes place for a variety of reasons, and whether a comprehensive refurbishment or a partial one is planned, the same basic principles discussed in this chapter will apply. The normal reasons that refurbishment takes place include general upgrading, changes and additions, replacement of old equipment or improving comfort, and the principles of sustainable refurbishment can be applied to, and included in, almost any action. When all of these interventions are added together, they can move the dwelling towards the goal of overall sustainability. Where a building is being completely refurbished, the opportunities normally exist to act on all areas and make the finished dwelling as sustainable as a new building built to the highest sustainable principles and practices. Where partial refurbishment is to be carried out, the work can contribute to an overall sustainability plan for the building and be a stepping stone towards the ultimate sustainability goal.

It should be borne in mind that the ultimate sustainability goal for housing is seen as 'zero carbon', normally defined as a dwelling that has a neutral import–export energy balance over a year measured in carbon emissions terms, and this can only be achieved by including renewable energy in the mix. This is discussed further in Chapter 9.

Technical Areas

Ten technical components of sustainable refurbishment are outlined in the following sections. It should be noted, however, that they are not necessarily separate elements; indeed, integration of actions in any building project is very important. The overarching principles are to reduce heat loss in winter and heat gain in summer that can cause overheating; to minimize all energy demands; to enable adequate ventilation; to reuse existing materials and incorporate new sustainable materials; and to reduce the consumption of water in the dwelling. Modelling of thermal aspects, energy use, overheating and condensation risk will always improve decision-making and confidence, and lead to a more integrated and efficient result.

Fuller details of actions in different situations and where to go for specific advice are given in Chapter 6.

Figure 1.1 External insulation being applied

Source: Simon Burton

Table 1.1 Improving the insulation of external elements

Component	Action
Walls	Solid walls can be externally or internally clad with insulation material depending on circumstances. Cavity walls can normally have their cavities filled with insulation.
Floors	Solid floors can be insulated by adding insulation above, with adequate flooring protection. Suspended floors can be insulated above or below.
Exposed ceilings	Where a loft exists, insulation can be added above the ceiling. Flat roofs can be insulated above the existing waterproofing using waterproof insulation, or below a new waterproof layer.
Windows	Double or triple glazing, with inert gas filling and heat-reflecting (low e) coatings, can replace existing windows and skylights, with thermally insulated frames.
Doors	Insulated doors can be used.
Cold bridging at junctions and from balconies	Cold bridging can be treated with careful detailing and extra external and internal insulation.

Improving the insulation of all external elements

Insulation has the combined effect of reducing the flow of heat through the element and reducing cold radiation effects on occupants, rendering lower air temperatures more comfortable. Equally, summer heat can be kept out by insulated external elements, although it is advisable to maintain adequate internal thermal mass in order to provide a buffer against overheating.

Ventilation that is adequate and efficient

During the heating season, controlled ventilation will minimize heat loss, while providing adequate indoor ventilation; during summer, more ventilation may be required to ventilate internal heat gains, although opening windows can normally be used.

Table 1.2 Adequate and efficient ventilation

Component	Action
Cracks and air paths	Cracks, open joints, service entries and unused chimneys can be sealed to reduce the uncontrolled movement of air.
Draught-stripping	All doors and windows can include draught-stripping to stop wind-induced air movements.
Natural ventilation	Trickle vents above windows and passive stack ventilation from kitchens and bathrooms can be used as a ventilation strategy.
Mechanical ventilation	Whole-house mechanical ventilation with very efficient heat recovery can be used.

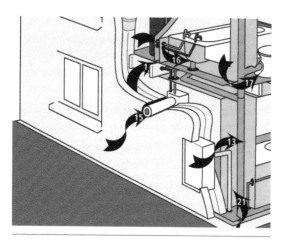

Figure 1.2 Air leakage routes are common

Source: Paul Jennings

Figure 1.3 Increasing passive solar gain

Source: Andy Simmonds

Providing efficient space heating

Heating is usually needed in all housing to warm up from a cold start, even though in very well-insulated dwellings internal gains can subsequently maintain adequate temperatures. Under-floor heating can be used, making lower air temperatures acceptable and enabling lower boiler flow and return temperatures to generate greater efficiency via more time spent in condensation mode.

Table 1.3 Providing efficient space heating

Component	Action
Passive solar gain	Opening up south-facing and other windows can provide useful solar heating.
Centralized or area heating	If available, linking to block or district heating systems, including those using biofuel and combined heat and power (CHP), is likely to be a sustainable option.
Gas or oil boiler	Efficient condensing boilers with optimized installation and control systems can provide efficient, controlled heating.
Heat pump	Electrically driven heat pumps (air or ground sourced) can provide efficient heating if they have high annual coefficients of performance and are well controlled.
Biomass heating	Enclosed biomass burners can provide efficient and sustainable heating since they rely on a renewable energy source.
Micro-combined heat and power (microCHP), cogeneration	Heat supplied from an individual microCHP system can provide a sustainable heating system where demand is high.

Table 1.4 Providing domestic hot water efficiently

Component	Action
Solar water heating	A well-designed, well-installed and well-controlled solar water system can provide around 50% of domestic hot water (DHW) use.
District or block heating system	If available, linking to block or district heating systems, including those with combined heat and power (CHP) is likely to be a sustainable option.
Gas and oil boilers	Efficient condensing boilers with optimized installation and control systems can provide efficient DHW production.
Heat pumps	Electrically driven heat pumps (air or ground sourced) can provide efficient DHW if they have high coefficients of performance and are well controlled.
Direct electric heaters	Small electric water heaters may be efficient for small and isolated DHW use.
Storage cylinders and pipework	Well-insulated storage cylinders, with high-performance heat-exchange coils, connected to the boiler by short insulated pipework, can maximize efficiency.
Reducing hot water use	Showers and reduced flow taps can be used to reduce hot water use.

Figure 1.4 Solar thermal collectors retrofitted during refurbishment

Source: Charlie Baker

Providing domestic hot water (DHW) by efficient means and controlling its use

In a low-energy house, domestic hot water energy use may be greater than space heating use unless care is taken.

Avoiding overheating that could require active cooling

Some refurbishment measures may increase the risk of overheating in hot weather, and sustainable refurbishment needs to guard against future demands for active cooling.

Table 1.5 Avoiding overheating

Component	Action
Reducing internal heat gains	Insulating hot water pipework and storage cylinders and using efficient electrical equipment will reduce internal and unwanted heat gains during summer.
Reducing external heat gains	Installing external solar shading on exposed east, south and west façades will reduce solar heat gain.
External landscaping	Trees can provide shading, and replacing hard surfaces with vegetation can reduce the temperature of air entering the house.
Use of thermal mass	Maintaining and using internal thermal mass can reduce temperature swings and enable cool night temperatures to reduce maximum day temperatures during summer.
Adequate ventilation	A combination of upper- and lower-opening windows or cross-ventilation can be used to ventilate out unwanted heat gains.

Figure 1.5 Permanent shading and vegetation for future shading

Source: Jes Mainwaring

Utilizing daylighting, efficient lighting and control systems

Table 1.6 Daylighting, efficient lighting and control systems

Component	Action
Optimizing daylight	Opening up, or providing new windows and skylights in internal dark spaces can reduce the use of artificial lighting.
Light surfaces	Light-coloured paint on internal surfaces will improve light levels due to better reflection.
Efficient lamps and luminaires	Tubular compact fluorescent lighting and light-emitting diodes (LEDs) in appropriate fittings can reduce electricity use.
Switching	Individual and convenient switching with dimmer controls, push buttons, and light intensity or movement sensor controls can minimize electricity use.

Installing efficient appliances and controls

In a low heat-demanding house, electrical appliances are likely to be the largest cause of carbon dioxide emissions.

Table 1.7 Installing efficient appliances and controls

Component	Action
Kitchen and utility equipment	Choosing the most efficient installed equipment will reduce electricity consumption, and gas cooking will reduce carbon emissions compared with most electric cooking.

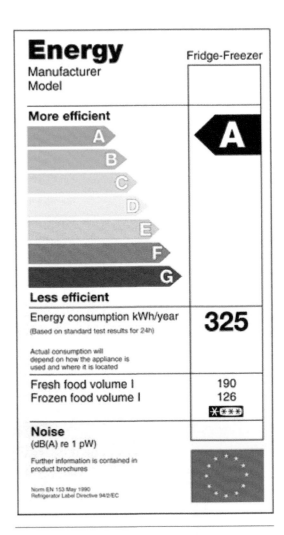

Figure 1.6 Domestic appliance energy label

Table 1.8 Minimizing water use

Component	Action
Toilets	Low-flush units can replace older systems.
Showers	Showers with low-flow aerated showerheads can be installed or replace baths.
Taps	Where taps are used for direct washing (rather than filling a basin), spray or restricted aerated flow can replace traditional taps.
Rainwater and greywater storage and use	Rainwater or greywater collection and storage can be installed to flush toilets and for watering the garden.

Installing equipment to minimize water use

Figure 1.7 Rainwater collection and storage

Source: Jeffery Pike

Reusing existing components and using new sustainable materials

Table 1.9 Existing components and new sustainable materials

Component	Action
Existing components	Reuse and restoration of a dwelling's components is likely to involve less carbon emissions than replacement.
New materials	New materials can be chosen with minimal embodied energy and environmental impacts in sourcing and manufacture.

The risks of sustainable refurbishment

Sustainable refurbishment can be accompanied by some risks that could cause subsequent problems related to condensation, damage to materials, freezing or overheating. Strategies to avoid these are a part of good design (see Chapter 6 for more detail).

Non-Technical Aspects

Sustainable refurbishment needs to take the whole process into account, from planning, design and implementation, through to follow-up during occupation and use.

Designing for the occupiers

Although the above actions can enable occupants to live a low-carbon and sustainable lifestyle, it is well known that it is the way in which occupants use their dwellings that determines actual energy use. Refurbishment can therefore be designed to enable and encourage sustainable behaviour, and to provide advice on how to do this (see Chapter 2 for further details).

Supporting the refurbishment process

Whether the refurbishment is carried out with the occupants in place or with an unoccupied building, the process of refurbishment can be designed and carried out to minimize local disruption from dust and noise by careful site management. Minimizing the waste of materials used on site, and separating and recycling waste are also important.

The UK Considerate Constructors Scheme is one method of encouraging good practice in the fields of environmental, workforce and general public concerns. Established in 1997, this initiative operates voluntary site and company codes of considerate practice, with which participating construction companies and sites register. The scheme is a non-profit independent organization founded by the industry to improve its image. All sites registered with the scheme are monitored by an experienced industry professional to assess their performance against the eight-point site code of considerate practice, which includes such categories as the environment, cleanliness and the importance of being considerate, a good neighbour, respectful, safe, responsible and accountable (see www.ccscheme.org.uk).

Figure 1.8 The Considerate Constructors Scheme

Comprehensive and partial refurbishment

Addressing all sustainability issues in a dwelling at one time could reduce energy use to new housing standards or even higher sustainable levels, and this is possible if the dwelling is being comprehensively refurbished, with adequate time and resources. However, this is unlikely to occur in much of the existing housing stock, and Chapter 3 looks at some strategies through which comprehensive sustainable housing may be achieved in the long term.

Occupation and feedback

While good design and construction should enable and encourage sustainable living, occupants need to be aware of how their house is performing and what they can do to optimize sustainability. Smart meters which give useful information on energy use, clearly marked controls that are located in easy-to-use places, and clear advice to new occupants are all areas where designers can support sustainable dwelling. This is the subject to which Chapter 2 turns.

2 Occupier Information and Behaviour Change

Occupant Behaviour

Sustainable refurbishment implies that houses should provide comfortable and low-carbon living space for many years, and this could well mean for several different occupiers over the years. Ideally, refurbishment should be designed with the needs and capabilities of the occupiers in mind to enable and encourage them to live sustainably. In reality, different occupiers have varying expectations and needs in terms of what temperatures are considered comfortable, hours of occupation, ventilation rates, equipment, and so on, and they also have different lifestyles. While most of the refurbishment measures included in this book are beneficial for all occupiers, it is worthwhile planning a refurbishment to meet any specific needs of the most common group of occupants.

Energy use in houses can vary according to a ratio of 3 to 1 depending upon the expectations and needs of different occupants in similar housing. Sustainable refurbishment will need to encourage occupants of all persuasions to operate their dwellings as sustainably as possible while satisfying their needs. There are a number of ways in which this can be done.

Exploring Occupant Needs and Behaviour

Questionnaires and surveys of existing occupants can focus on expectations so that occupants feel their concerns and needs are being taken into account and are thus more likely to identify with changes, adopt new behaviour, when necessary, and support the principles of sustainable refurbishment. Surveying housing-related difficulties or problems experienced by occupants can give an understanding of what heating and ventilation equipment and control systems will best suit future occupants. Where housing is to be occupied by new occupants, it may be useful to

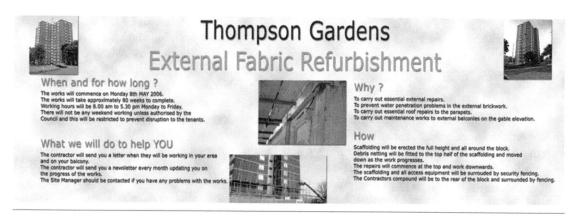

Figure 2.1 Information for tenants

Source: Sandwell MBC

discover the likely family and occupation profile of future occupants. Young couples, for example, are likely to live different lifestyles than families and the elderly, and thus may wish to use their dwellings in different ways. Working with groups of future residents to explore issues and to develop sustainable solutions is likely to be an effective solution.

Designing to Suit Occupant Lifestyle

Sustainable features, such as insulation, efficient lighting, the reduction of summer heat gains, efficient appliances, and the use of sustainable materials, will bring universal benefits, while some areas will benefit from considering their likely use by different occupants. For example, whole-house ventilation systems are particularly appropriate for smaller dwellings occupied by families; combination boilers may be most suitable for occupants with small domestic hot water (DHW) needs; solar water heating may prove more efficient for those who have daytime DHW demand; micro-combined heat and power (microCHP) will only be effective for large houses with continuous occupation; and wood-burning stoves will only be usable by active occupants prepared for refuelling and ash removal.

Avoiding Complicated Control Systems

Although some occupants many take a particular interest in understanding and operating complicated control systems, most will not be interested in or wish to understand the intricacies of a control system for heating or other energy-using functions. Normally, a control system that gives the occupant a limited number of options is best, making it easy for that occupant to use an efficient setting. Complicated control systems may simply encourage the occupant to override automated controls and rely on manual control. Control systems with efficient default settings (e.g. standard heating periods and temperatures) or push-button limited-time overrides to boost space heating or DHW will provide the most efficient fall-back operation. Well-located controls with easily understandable visual displays are important.

Providing Tailored Information

Information with varying levels of detail will suit different users and can encourage sustainable occupation. Traditional information has focused on equipment safety concerns and comfort; but an emphasis on sustainability can be provided by reinforcing the potential for reduced fuel bills and, in some cases, a lower household carbon footprint. Two levels of information are appropriate: the quick 'getting-you-going' guide and a fuller explanation of systems and how to operate them most efficiently. Both should contain information on time and temperature settings for space heating and DHW, on ventilation (requirements or settings), on lighting and appliance choices, and on any specific installations, such as solar water heating, which may be unknown to the occupants. Where possible, information should be fixed adjacent to the equipment to which it relates.

In the more detailed guidebook, additional information can include the following:

- a tabulated range of typical fuel consumption for different times and types of occupation;
- a description of all the energy-saving measures included and how they work;
- additional energy-saving measures that could further reduce consumption;
- opportunities for lifestyle changes to reduce one's carbon footprint.

Guidance on communicating with occupants is given in the UK Energy Saving Trust publication *Reducing Emissions from Social Housing* – in particular, 'Technical Annex Section E: Encouraging behavioural change' (see www.energysavingtrust.org.uk/business/Publication-Download/?oid=1529508&aid=4091160).

Installing and Setting Up a Smart Metering System Providing Information on Energy Use to Occupants

Smart meters that display information on energy use so that householders can understand and reduce the amount of energy that they use are rapidly developing. The range of information displayed could theoretically include historic consumption, target consumption, current consumption, alarms, cost conversion, and weather, etc. Information on the use of smart meters and interpretation of the data provided is necessary to make the meter effective.

Setting Up a Communal Information Exchange and Problem-Solving System

Word of mouth and local discussion can be the best way to support sustainable living when refurbishment has taken place and occupants need to change their habits to optimize the benefits provided. A local person can be briefed about energy-saving measures and act as a focal point for information dissemination and problem-solving. A trusted local person on a housing estate can be more effective at distributing information than a published guidebook, and may be more effective at persuading occupants of the benefits of using their houses sustainably.

Behaviour Change

However energy efficient a house may be, it is the occupants who will determine actual energy use. Behaviour or lifestyle modification is possible with regards to room temperatures, the number of rooms used, clothes worn, duration of heating, ventilation, use of electrical equipment, washing, lighting arrangements and many other factors. This has happened in specific

Figure 2.2 Occupants discuss sustainable living

Source: Hockerton Housing Project

situations where residents have come together or have been selected as good examples of sustainable living, such as in the Hockerton Housing Project. While this is a desirable long-term goal, immediate steps should be to refurbish dwellings in order to promote and facilitate lower carbon lifestyles, and to provide information and support for long-term behaviour change.

3 Strategies to Get Action for Sustainable Housing

Worldwide, sustainable new housing is driven by legislation, with building codes and regulations requiring the incorporation of minimum energy standards. These are enforceable by law and can be upgraded over the years – for example, in the UK, all new housing must be 'carbon neutral' after 2016. For the existing housing stock, a different set of mechanisms is required to upgrade sustainability, and there is a need to exploit every opportunity and to provide encouragement in a variety of ways.

There are many reasons why existing housing stock is more difficult to address than the new-build sector, including issues of ownership and tenure, disruption, and variations in age, type, construction, state of repair and use. Blanket approaches covering areas or estates may be possible, particularly in the rented sector, and legislation can be used to a limited extent across the board. However, neither of these is likely to tackle the scale of the problem within the timeframe that governments need to satisfy their policy objectives of making the existing housing stock sustainable.

It is important, however, to remember that sustainable refurbishment has enormous tangible benefits for occupants, and it is these benefits that occupants frequently describe with regards to their sustainable house, as reported in the case studies presented in this volume (see Chapters 4 and 5). Occupants of old housing frequently complain about a cold house, draughts, condensation and mould growth, ambient temperature slow to warm up, large fuel bills, and poor ventilation. Occupants of sustainably refurbished dwellings eulogize about a warm house, warm walls and floors, the rare need to turn on the heating, ambient temperature quick to warm up and slow to cool down, good air quality, and very low fuel bills – in general, a very comfortable house and a house of quality. Owner-occupiers frequently spend large amounts of money on refurbishment to produce a home of quality and comfort. Those who have carried out a sustainable refurbishment are likely to find that they have gone a long way towards meeting this goal. The providers of rented housing are likely to get very similar reactions from their tenants. Those providing advice and design on sustainable refurbishment may find that the enticements of comfort and quality in a home are more powerful than those of sustainability.

How Sustainable Refurbishment Happens

In most countries, the existing housing stock does not stay untouched for years, and it is common to have work carried out at regular intervals, both as necessary maintenance and for improvements, replacements and extensions. These activities frequently provide an opportunity to enhance energy performance. There are also other trigger points that can be used, as well as financial (and other) stimulations. The following are the five most common opportunities and intervention points for sustainable refurbishment.

Comprehensive refurbishment

When a building is stripped back to its basic structure for comprehensive upgrading, all sustainability issues can be relatively easily addressed; the building can be made energy efficient; and this is likely to be required by building controls. This is more common in the social housing sector or where an area of derelict housing is being refurbished, or a building converted to housing by a single owner. Owner-occupied housing is more likely to be incrementally refurbished except when a property changes hands; in this case, comprehensive refurbishment occasionally takes place. It is important that high standards of sustainability are included in all comprehensive refurbishments as the opportunity will not be repeated for many years, and this is by far the easiest and most cost-effective time to adapt the building to future demands. When comprehensive changes take place, the barriers of disruption and availability of finance have at least to some extent been overcome and small additions to the works will be more acceptable. Funding organizations are likely to be more receptive to additions that extend the life, value and sustainability of the dwelling. Later sections of this chapter discuss legislation and the effects of implementing the new European Union directive.

Opportunistic approach

When any sort of repair or replacement is needed and disruption is bound to occur, it may be worth carrying out additional sustainable

Figure 3.1 Comprehensive refurbishment

Source: A2M

upgrades. New kitchens and bathrooms, loft conversions, window, roof or boiler replacements, damp-proofing, rewiring, redecoration – all present ideal opportunities to introduce energy-saving measures. One frequent objection to installing energy-saving measures is the disruption to household life and the dirt generated. This is also the time, however, when finance is found or mortgages increased, and the marginal extra cost can thus be more easily accommodated. Most of these repairs or replacements can be accompanied by wall or roof insulation, installing high-performance windows, and using A★-rated equipment, all of which can enhance the original project, as well as rendering a building more sustainable. Both designers and contractors are in a good position to promote and ensure the implementation of sustainability

measures during the execution of such work. Just as a householder might have a plan to upgrade different elements of their house as money becomes available, these interventions should form part of a whole-house sustainability plan, as described later in this chapter.

New purchase or new tenancy

When a building's occupant changes, this is often a trigger for physical changes in the dwelling. Often there is a period of time when the building is unoccupied and when works can be undertaken without disruption; indeed, the new occupant may well require or expect some modifications to the dwelling to suit their needs. This is also the opportunity for legislative requirements to be carried out – for example,

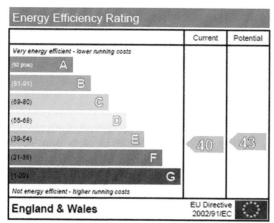

Figure 3.2 Energy performance certificate

the requirement for the owner to carry out an energy audit and present an energy certificate to the new owner or tenant. Certificates can include likely fuel bills, substandard features and potential improvements which can influence the purchaser or renter and have an impact upon the value of the dwelling. Governments can also set minimum energy standards or equipment ratings when there is a change in ownership or tenancy much more easily than attempting to legislate for all occupied dwellings. New rents may be set or higher mortgages raised so that the requirements for higher energy measures or standards and the financial benefits can be brought into the cost equation, and owners, tenants and finance providers can work to a common sustainability goal. The change in occupation is the time when market forces, in terms of costs and benefits, come into play; comfort and fuel costs will play an increasing part in this process. This important opportunity needs to be recognized by property agents and social housing organizations, and appropriate time reserved for the design and implementation of additional sustainability works.

The green owner

There are a growing number of organizations and individuals who have a specific desire or commitment to make homes more sustainable – mostly providers of social housing or committed owner-occupiers. Despite this important starting point, it may have been difficult in the past to find designers and builders with the knowledge and experience necessary to undertake sustainable refurbishment at an acceptable cost. In tandem with the growing general publicity about the need for sustainability is a wealth of knowledge of the practical applications involved; there is also a market advantage for professionals to sell themselves on this basis. The green owner needs to select architects, consultants, builders and installers who can demonstrate their knowledge and experience of sustainability in their area. There are a number of ways in which this

can be achieved – for example, by checking out organizations with friends and colleagues; by asking perceptive questions of tendering organizations; by visiting previously carried out work; and by obtaining references and relevant training experiences of the organizations. It is also becoming far easier to acquire lists of builders registered by national or local government who can carry out specific works, such as external insulation and installing renewable energy sources.

Area improvements

Just as new houses are rarely built as individual units, and streets or groups of dwellings are much cheaper to construct, so there are numerous benefits when refurbishments are carried out on a geographical area basis. Neighbours can see improvements taking place and can visit and discuss issues with others in similar positions. There are benefits to the locality by house improvements contributing to the general upgrading of the street and, consequently, raising house values. Builders can benefit from having several sites close together and carrying out similar works to similar properties (e.g. it may be possible to carry out the same works, such as external insulation or re-roofing, to a series of adjacent houses). Group purchasing of materials and combined works can result in reduced costs. Owner-occupied and rented dwellings can be treated in similar ways, and the general level of activity may enable pressure to be put on landlords who would not otherwise see the benefits of sustainability improvements. Local authorities can focus resources such as information, financial support and building control more efficiently on a specific area. Adjacent areas may observe that other neighbourhoods are being improved and may see the advantages of similar actions.

Pump-priming actions by local authorities (or even residents' associations) have been used in the past to upgrade areas of existing housing in order to provide necessary improvements and facilities. More recently, this has also extended to

Figure 3.3 Owners are pleased with their sustainable refurbishments

Source: Tamzin Doggart

establishing new levels of sustainability. Currently, projects such as Housing Renewal Areas and the Greater London Authority's (GLA's) Energy Action Areas demonstrate area-focused activities.

Drivers and Enablers

Legislation

Legislation, although perhaps not popular with existing homeowners, is likely to be an increasingly important driver in making existing housing stock more sustainable. The proposed recast of the European Union's Energy Performance of Buildings Directive (EPBD) states:

Major renovations of existing buildings, regardless of their size, above a certain size should be regarded as providing an opportunity to take cost-effective measures to enhance energy performance. Major renovations are cases such as those where the total cost of the renovation related to the building shell and/or energy installations, such as heating, hot water supply, air-conditioning, ventilation and lighting, is higher than 25 per cent of the value of the building, excluding the value of the land upon which the building is situated, or those where more than 25 per cent of the building shell undergoes renovation. For reasons of cost efficiency, it should be possible to limit the minimum energy performance requirements to the renovated parts that are most relevant for the energy performance of the building.

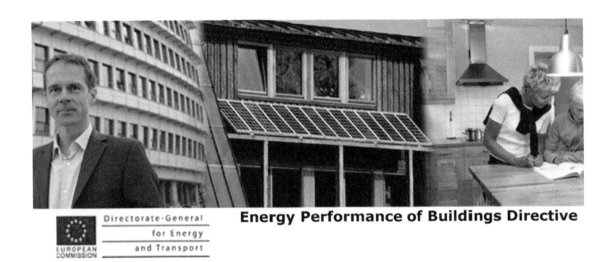

Directorate-General
for Energy
and Transport

EUROPEAN
COMMISSION

Energy Performance of Buildings Directive

Figure 3.4 Proposed revision of the Energy Performance of Buildings Directive (EPBD)

The transposition of the EPBD into the legislation of EU member states will have a large effect on major renovations in the future. Requirements to provide energy certificates and minimum standards when a dwelling changes ownership or tenancy are already common in the EU due to the requirements of the first EPBD (see earlier in this chapter). Where elements of a building are replaced, such as windows, boilers and walls, these are commonly required to be at the standard required for new construction. 'Consequential improvements', as required by UK building regulations for non-domestic buildings, are being considered for housing. This can require that when money is spent on general improvements, extensions or conversions to a building, a proportionate amount of money must be spent on sustainability improvements to the building in general.

Cars are commonly required to have a safety and efficiency check each year, and this principle could be applied to dwellings, with those falling below set standards needing to be upgraded if they are not to become 'unsuitable for human habitation', to use a UK phrase relating to health and safety issues.

Financial support

Finding finance or an appropriate financing mechanism is frequently a barrier to sustainable refurbishment, and the availability and knowledge of funding options is a prime enabler for owner-occupiers and block owners and landlords.

At the European Union level, structural funds – in particular, the European Regional Development Fund (ERDF) – can be used to help finance national programmes and projects aimed at improving energy efficiency in the residential sector. These are accessible via national governments.

Several types of national financial support and financial mechanisms are available for sustainable refurbishment. Examples include the following.

Grants:

• The Danish Energy Authority has a scheme to give grants for energy-saving measures in pensioners' dwellings.

- An Italian scheme allows a fiscal deduction of 55 per cent of expenses for works directed at improving existing building performance, spread over five years and available for people and companies.
- The UK Carbon Emissions Reduction Target (CERT) scheme requires energy suppliers to provide grants, and offers to help pay for energy-efficiency measures and renewable energy technologies for homes. The scheme is funded by a levy on fuel bills.

Preferential loans:

- In Germany, the CO_2 Building Rehabilitation Programme 61 (Energy-Efficient Construction and Rehabilitation) supports extensive energy rehabilitation measures in residential buildings completed in 1983 or earlier. The programme applies to owners of single-family or two-family houses or private apartments in homeownership associations or housing companies. Financial support is provided by loans at a fixed interest rate for ten years.
- In the UK, the Home Energy Pay As You Save (PAYS) scheme is being tested using pilot projects launched in December 2009. These pilots are testing consumer interest in PAYS, which offers householders capital to meet the upfront costs of installing energy-efficiency and renewable energy measures to existing homes. Householders then repay this finance through instalments that are lower than their energy bill savings.
- Some UK building societies will give additional loans for sustainable homes and sustainability improvements.

Market-based instruments:

- The French have a system called White Certificates. These certificates can be created from projects that result in energy savings beyond business as usual by target market actors or by energy service companies.

Certificates received for savings achieved by market actors can be used for their own target compliance or can be sold to other parties.
- The Czech Republic has a Green Investment Scheme that recycles money received through the sale of governmental emissions rights as grants for retrofitting improved energy efficiency in buildings.

Energy service companies (ESCOs):

- An ESCO is an organization that delivers energy services and/or other energy-efficiency improvement measures in a user's facility or premises, and accepts some degree of financial risk in doing so. The payment for services delivered is based (either wholly or in part) on the achievement of energy-efficiency improvements and on the meeting of other agreed performance criteria.
- In Germany, the city of Berlin, in partnership with several ESCOs, pays upfront for retrofits and the building owners pay the amount back over an agreed period, usually 8 to 12 years in annual instalments from the energy-cost savings.

Special government grants for capital improvements, such as insulation and boiler replacements for the elderly, people on benefits and those in fuel poverty, are also common. Local authorities in several countries provide additional funding stimuli for the sustainable refurbishment of houses in their area.

Figure 3.5 UK Ecology Building Society: 'A building society dedicated to improving the environment by supporting and promoting ecological building practices and sustainable communities'

Active promotion of available finance for sustainable refurbishment can be an important stimulus to action. The Energy Saving Trust in the UK maintains an active grant-searching facility on its website covering grants from all sources.

A whole-house sustainability plan

A sustainability plan can be useful to encourage integrated and comprehensive refurbishment of a dwelling where improvements are likely to take place over a number of years. This is based on a survey of the house carried out by a qualified professional who provides an easy to understand report to the owner-occupier. The report details all of the different elements of the house that could be upgraded, and quantifies the benefits of these different upgrades expressed in terms of cash saved versus cash spent, and energy/CO_2 saved. It also shows how costs could be reduced by combining certain upgrades with others (e.g. rewiring with internal insulation for external walls). The purpose of the plan is to give home-owners a route map for the next, say, 20 years of the improvements that they could make as and when the funds become available, or what to include when they conduct other upgrades. The sustainability house plan is best integrated with requirements to provide information on the energy performance of the dwelling when it is sold or re-let, and will clearly enhance the understanding of the new owner or tenant in terms of what has been done and what will be needed in the future.

Designers and tradespeople

House owners normally need to rely on the knowledge and skills of their designers and building contractors, particularly in terms of sustainability issues, unless they themselves are experts. Owner-occupiers frequently rely on builders they know and have used previously, or who are recommended by their contacts and friends, to outline what needs to be done and what systems and products should be installed. Tradespeople trained and certified in sustainable refurbishment can suggest sustainable measures to householders in terms of both comprehensive and partial changes, and should be able to implement them satisfactorily. Certification is also important so that the owner can rely on the information and recommendations put forward by the builder, and not suspect that the builder is simply trying to increase the scope and, therefore, the cost of the works. Knowledgeable and experienced architects, engineers and cost consultants are equally important when they are employed for a refurbishment project.

Show houses

Householders, both owners and tenants, contemplating improvements to their dwellings are likely to want to see, touch, discuss and experience sustainability improvements that others in similar situations have carried out. Show houses in a locality can stimulate and encourage others to follow suit, and allay fears, and can be the houses of owner-occupiers wishing to spread the word, a local authority for general publicity, or social landlords to consult with and inform their tenants. The UK Sustainable Energy Academy has set up the Old Home SuperHome scheme, a network of private houses refurbished to provide 60 per cent+ reduction in carbon emissions, providing details on their website and open days. Several UK local authorities have demonstration houses which have been refurbished to high sustainability standards and are also open to the public.

Figure 3.6 The Old Home SuperHome network, created by the UK Sustainable Energy Academy, is made up of older homes across the country that have been updated to improve their energy efficiency

4 Good and Best Practice in Single-Family Houses

Most single-family housing is owner occupied, but may be privately rented or owned by social housing organizations. The drivers for and process of refurbishment may be different; but the principles and physical details will be the same.

There are many reasons why work on an individual house takes place – for example, to repair faults, for general maintenance, to make improvements, to stop rising damp, to adapt a newly acquired house, to convert a loft or other space, or just to replace an antiquated kitchen or bathroom. While these all provide the opportunity to make the dwelling more sustainable, there are many reasons why owners may not undertake comprehensive refurbishment or may not include sustainability, including the following:

- There may be little awareness of the need or the opportunities.
- Sufficient finance may not be available.
- It may not be possible or desirable to change the external appearance – for example, to apply external insulation (e.g. due to conservation area status).
- It may not be desirable to reduce the size of internal spaces or remove features such as ceiling mouldings.
- The disruption of comprehensive refurbishment may be unacceptable.
- Only certain areas may be planned for refurbishment (e.g. a new kitchen).

The owners and designers in the following case studies have overcome these difficulties and gone as far towards comprehensive sustainability as was feasible at the time of refurbishment, taking into account the limitations listed above. While there may be technical limitations in specific situations, there are generally technological solutions to cover all aspects of sustainable refurbishment, and research and development of better methods and products are always ongoing. Frequently, there are a range of solutions to improve the sustainability of different elements applicable to different situations, as well as a range of standards that can be attained, from minor improvements to near zero energy loss.

The German *PassivHaus* concept, designed originally for new housing, includes very high standards and these are also being used for the refurbishment of existing housing. The *PassivHaus* concept is a comprehensive package of high insulation levels of all elements, as well as controlled whole-house ventilation, which results in minimal space heating requirements and can be seen as the ultimate goal for sustainable refurbishment if combined with appropriate renewable energy sources, efficient appliances and sustainable management by the occupant.

The case studies in this chapter represent good and best practice across a range of situations, and adopt a range of solutions. The selection, however, does not represent the results of a cross-country survey; rather, these examples have been chosen to demonstrate that hundreds or thousands of excellent examples of sustainable refurbishment already exist and that many owners have explored options and been able to achieve high standards, saving 70 per cent or more of energy consumed compared with much of the current housing stock.

Table 4.1 The case studies

Case study number	Type	Location	Age	Restrictions	Wall insulation to existing walls	Roof insulation	Floor insulation	Ventilation	Windows	Heating system	Renewables	Other
	Case studies				Sustainability solutions							
4.1	Detached house	France: Caluire, Rhone	1950	Some	External: 200mm cellulose plus 60mm wood fibre	350mm cellulose	10–150mm cellulose	Whole-house mechanical with heat recovery 90% efficient	Triple glazed, argon filled	Wood pellet stove; also heats DHW and bathroom radiators	Solar water heating: 6m²; and PV array: 16.3m²	Designed to comply with *PassivHaus* standards
4.2	Semi-detached house	France: Bourg-en-Bresse	Early 1900s		Internal: 100mm of wood fibre and 50mm of cellulose fibre	200mm wood fibre and 60mm cellulose fibre	100mm wood fibre and 50mm cellulose fibre	Whole-house mechanical with heat recovery 90% efficient	Front secondary glazed, rear double glazed, argon filled	Gas-fired central heating plus wood stove	Solar water heating: 4m²	
4.3	Semi-detached house	UK: Hereford, Wales	1869	Few	External: 250mm polystyrene/ 350mm mineral wool on east façade	400mm glass fibre	100mm polyurethane/ 250mm polystyrene/ 225mm sheep's wool	Whole-house mechanical with heat recovery 92% efficient	Triple glazed, argon filled	Gas-fired boiler	Solar ready	Aimed at *PassivHaus* standards
4.4	Terraced house, three to four storeys	UK: London (Culford Road)	1830	In a conservation area	Internal: 140mm glass wool plus 30mm air gap	180mm glass wool	100mm extruded polystyrene on new solid floor	Whole-house mechanical with heat recovery 92% efficient	Sash windows 4mm argon-filled double glazing with low e; triple glazed at rear	Gas-fired boiler; under-floor heating on ground floor plus towel rails	1.32kW peak PV array	
4.5	Semi-detached house	UK: London (Carshalton Grove)	1870	None	Internal insulation using five different systems	200mm cellulose fibre in flat and sloping areas	New solid floors: 300mm polystyrene Existing wood floors: 175mm cellulose fibre	Natural with window trickle vents and heat recovery extract fans	Front sash double glazed with argon fill and low e; triple-glazed window at rear; double-glazed Velux roof lights	Gas-fired under-floor heating in whole house	4m² of evacuated tube solar collectors for hot water	Rainwater collection for toilet flushing
4.6	Semi-detached house	UK: London (Chester Road)	Late 19th century	In a conservation area	Internal insulation: 100mm and 50mm fibreboard	150mm mineral wool plus 75mm wood fibreboard, and 250mm mineral wool on extension	Under floor: 150mm mineral wool plus 20mm wood fibre New floor: 100mm polystyrene	Natural with draught-proofing and heat-recovery extract fans	Double glazed with argon fill and low e, as well as draught-proofing	Gas-fired boiler plus wood stove	2m² of solar panels for hot water; 10m² of PV panels	Low-energy lighting, AAA-rated equipment, water saving and recycling
4.7	Detached house	UK: Bristol	1930s	New dormer windows required permission	External insulation on all four façades; 80mm polyisoçyanurate	100mm mineral wool between rafters; 70mm overlay board on top	150mm cellulose insulation between joists	Natural with trickle vents and an extract fan	Double glazed, low e coated, argon filled in timber frames	Gas-fired boiler	4m² solar thermal for DHW; 1.225kW peak PV panels	Low-energy lighting, water saving, use of recycled and natural materials

	Type	Location	Year	Planning issues	Wall insulation	Roof insulation	Floor insulation	Ventilation	Glazing	Heating	Solar	Other
4.8	Semi-detached house	UK: Manchester	1850s	Wood gasifying boiler required special permission	Internal insulation on front and back walls: 27mm Aerogel, external on gable end, 170mm wood-fibre rendered	Flat roof: 350mm glass wool, pitched 200mm sheep's wool	200mm glass wool under floors	Natural	Existing: double glazed with argon fill New: triple glazed with argon fill	Gasifying wood boiler plus solar thermal	10m² evacuated tube collectors for space and water heating	Low-energy lighting, water saving, natural materials
4.9	Detached country house	Portugal: Western Algarve	1970s	Protected rural environment	External insulation: 60mm expanded polystyrene and thin-coat render; extends 700mm below ground	New and existing roofs: 110mm expanded polystyrene	New tiled concrete floors: 200mm Leca expanded clay insulated screed	Natural, trickle vents over windows and presence-detection extract fans	New larch double-sealed, double-glazed windows and glazed doors; argon filled, low e	Propane gas boiler for space heating and solar domestic hot water backup	Solar thermal collectors for hot water	Emphasis on passive solar gains introduced with the refurbishment
4.10	Detached house	US: Oakland, California	1915	Yes	Internal: cellulose	Cellulose	Cellulose under floor	Natural with extract fans	Double glazed with low e coatings	Gas-fired low-temperature boiler	Solar thermal and 2.72kW peak PV panels	Reuse of materials, water-saving devices and Energy Star appliances
4.11	Terraced house, three storeys	Germany: Cologne	1929		External insulation: 160mm	240mm	80mm	Whole-house mechanical with heat recovery	Double glazed	Gas-fired boiler with solar panels	4.6m² of solar thermal collectors	
4.12	Detached house	Germany: Oldenburg	1869		External insulation: 160mm	280mm cellulose	30mm vacuum insulation panels in basement	Whole-house mechanical with heat recovery 92% efficient	Triple-glazed PassivHaus standard	Wood pellet boiler with solar collectors	8m² of solar panels	
4.13	Detached half-timbered house	Germany: Eichstetten	1750	Listed building	160mm cellulose insulation injected into cavities and 60mm–80mm wood fibre/cellulose boards	200mm cellulose		Whole-house mechanical with heat recovery >90% efficient	Single glazed with new seals	Wood pellet boiler plus solar collectors	13m² of solar panels	
4.14	Detached country house	Germany: Constance	1959	None	300mm cellulose inside a new timber façade	300mm cellulose-blown fibre	300mm polyurethane foam	Whole-house mechanical with heat recovery >80% efficient	Triple glazed	Air-to-water heat pump plus solar collectors	4.1m² of solar panels	
4.15	Central city top-floor flat	UK: London (Kings Cross)	1905	No planning issues, but seventh floor	Internal insulation: 70mm polyisocyanurate rigid board	Internal 55mm polyisocyanurate and phenolic foam	None	Natural, with extract fan in bathroom	Secondary and double glazing	Gas condensing combination boiler	None	

Case Study 4.1 France: Caluire, Rhone

- 1950s solid-walled, detached house converted to near *PassivHaus* standards.
- External insulation to all walls.
- Extensive architectural changes.
- Solar water heating.
- 16.3m² of photovoltaic panels effectively make the house carbon neutral.
- Total costs of acquisition and refurbishment were within the market value of houses in the area.

Figure 4.2 South façade after renovation

Source: Jean-Philippe Claudel, Pierre Lévy and Sophie Zélé

Figure 4.1 Original south façade

Source: Jean-Philippe Claudel, Pierre Lévy and Sophie Zélé

Description

This large detached house is a few kilometres from the centre of Lyons, France, but in a green and sunny environment in contrast to the centre of Lyons. The house was built in 1953, and constructed with a basement of reinforced concrete and upper walls of clinker concrete. The floors are constructed from reinforced concrete beams with brick coffering in between. The windows were single glazed with wooden shutters, and the heating was from an oil-fired boiler in the basement. A major refurbishment was required, with significant additions and changes to the architecture, aiming for *PassivHaus* standards.

Refurbishment Process

After purchasing the house, the owners consulted several architects to choose one who was capable of redesigning the house to *PassivHaus* standards. After extensive studies of the house and the local surroundings, refurbishment started in 2007 and finished in 2008. The architectural changes were to open up the house to the south and to protect the north elevation. A new entrance space was designed on the north side and a terrace on the south garden side. A further major change was the external insulation applied to all façades, which effectively removed the original architectural features and replaced them with a new appearance. Other important work was to dismantle the chimneys in the loft in order to reduce thermal bridging. Lower down, the chimneys were used as service ducts.

Figure 4.3 External timber framing

Source: Jean-Philippe Claudel, Pierre Lévy and Sophie Zélé

Sustainability Improvements

Table 4.2 Sustainability improvements in Case Study 4.1

Walls	External insulation to all walls; 200mm timber frame filled with cellulose fibre and a cover of 6mm of wood fibre and external coating.	U value: 0.15W/m²K
Floors	100mm–150mm of cellulose fibre under the existing floor. The walls of the stairs to the basement were also insulated.	U value: 0.25W/m²K
Roof	350mm of cellulose fibre.	U value: 0.1W/m²K
Windows	New triple-glazed windows with argon fill. Wooden solar-shading devices were added on the south side.	U value: 0.8W/m²K
External doors	No changes were made.	
Other sustainability features	• Rainwater is collected in basement tanks and double filtered for consumption. • Water-based paints and varnish were used. • Dry toilets were installed on the ground floor and were linked to a compost system.	
Airtightness	Joint sealing was undertaken with all works.	Blower door tested air change rate: 1.2ACH (further sealing has been carried out)
Ventilation	Whole-house mechanical ventilation system with heat recovery.	Heat recovery: 90%
Heating	Pellet wood-burning stove in sitting room.	
Renewable energy sources	• 6m² of solar thermal collectors for domestic hot water. • 16.3m² of PV panels.	

Figure 4.4 Finished wall after external insulation

Source: Jean-Philippe Claudel, Pierre Lévy and Sophie Zélé

Comfort and Overheating

Having initially decided not to install a heating system, a wood-burning stove was considered a necessary addition to the sitting room after the first winter. This has provided adequate heating for the occupants. The combined effect of shading over the south-facing windows and the large trees on the south side has prevented any summer overheating.

Change in Energy Use and CO_2 Emissions

Costs

The total cost of the refurbishment was more than 300,000 Euros; but most of this was general refurbishment and improvements to the dwelling. Taken with the purchase cost of the house, the total expenditure is within the market price for a house in this area. However, the money was spent on the insulation and the architectural changes, rather than on a swimming pool or a luxurious kitchen or bathrooms. Since this was one of the first *PassivHaus* refurbishments in France, the cost of some of the technology, along with a difficult access, raised the overall project costs.

Planning Issues

The house was situated less than 500m from a building protected by Historic Monuments of France and was visible from it; as a result, negotiations were undertaken for the extension of the house. A construction permit was necessary in order to examine the stability of the ground and the house, and to install solar panels (which, at the time, were normally refused without an architectural study).

Table 4.3 Change in energy use before and after refurbishment in Case Study 4.1

	Before refurbishment	After refurbishment	Savings
Wood used	Unknown	4681kWh/year	Unknown
Electricity demand, including equipment	Unknown	2439kWh/year	Unknown
PV electricity generated	0	1762kWh/year	
Total fuel costs	Unknown	413 Euros per year repayment to owners, taking into account feed-in tariff	

Conclusions

A heat pump based on exhaust air was considered for space heating, but was rejected, and a wood stove was preferred for environmental reasons. A wood pellet cooker with a boiler was also considered for hot water during the winter and for bathroom radiators, but was not installed. Future work might include insulating the basement exterior walls to reduce cold bridging above.

This project demonstrates that it is possible to renovate a house to near *PassivHaus* standards with a reasonable budget when other works need to be carried out.

Figure 4.5 View from the terrace

Source: Jean-Philippe Claudel, Pierre Lévy and Sophie Zélé

Case Study 4.2 France: Bourg-en-Bresse

- Large semi-detached urban house, completely reorganized internally.
- Full internal insulation in order to retain external appearance.
- Low-temperature gas-fired heating, plus wood stove.
- Solar water heating.
- Whole-house ventilation.

Figure 4.6 The appearance of the house from the road has not changed

Source: Myriam Jacquet and Gregoire Magnien

Description

This 200m² semi-detached house situated in the suburbs of Bourg-en-Bresse, in eastern-central France near Switzerland, consists of two storeys above a ground floor, including the garage. The house, built during the 1930s, was constructed with solid concrete walls on the ground floor, with clinker construction above, approximately 480mm thick (a common construction in this region). The house was in good condition with a recently replaced, though almost uninsulated, roof; and the idea was to main-

tain the external appearance of the house and totally change the internal arrangement – hence, the internal insulation of the walls and roof. Before buying the house, the owners would have preferred a house with a south aspect, but settled for this house due to its excellent location and the great potential for upgrading. They also believed that an east–west orientation had many advantages and that the house offered the opportunity to produce a very low-energy house, including the large south-facing roof, which was perfect for solar panels.

Refurbishment Process

The house was bought by the architect owners in 2006 and the refurbishment began in 2007, with a second phase completed in 2010. The house was unoccupied during the first phase of works. The front façade was left unchanged, retaining all the original detailing, while the internal partitions were all removed and a new layout produced at the same time as insulating the external walls. Installation of solar panels on the south-facing roof was based on a short study to avoid shading from the roof of the adjacent house and the chimneys.

Figure 4.7
Secondary glazing is used on the front windows in winter

Source: Myriam Jacquet and Gregoire Magnien

Sustainability Improvements

Table 4.4 Sustainability improvements in Case Study 4.2

Walls	Internal insulation between metal framing: 100mm of wood fibre and 50mm of cellulose fibre.	U value: 0.24W/m²K
Floors	Insulated below suspended timber floor with 100mm wood wool and 50mm of cellulose fibre.	U value: 0.23W/m²K
Roof (sloping)	Insulated between new rafters with 200mm of wood wool and 60mm of cellulose fibre.	U value: 0.15W/m²K
Windows	• Front: new double glazing inside the existing windows, removed in summer. • Rear: aluminium frames with thermal break, double glazed with low emissivity coating and argon fill.	U value: 0.14W/m²K
External doors	Front door not changed.	
Airtightness	A vapour barrier was placed between the layers of insulation on the walls and roof. No airtightness tests were carried out.	
Ventilation	Heat-recovery ventilation system (Aldes group double-flux Dee Fly)	Heat recovery: 90%
Space heating	Wall-mounted condensing gas boiler with boost fan and wood stove in the living room.	
Domestic hot water	Solar panels with gas (priority) or electric backup.	
Renewable energy sources	Solar water heating comprising 4m² Velux panels; 20m² of photovoltaic added in 2010.	60% of annual DHW demand

Figure 4.8 Steel framing to receive internal insulation

Source: Myriam Jacquet and Gregoire Magnien

Comfort and Overheating

Before refurbishment, the top floor of the house was known to overheat in summer due to the virtual lack of insulation. The ventilation is good and aqueous paints were used, and the wood treated with environmentally friendly products. There are no complaints about overheating since refurbishment, although no air conditioning was installed.

Change in Energy Use and CO$_2$ Emissions

Table 4.5 Change in energy use before and after refurbishment in Case Study 4.2

	Before refurbishment (calculated)	After refurbishment (calculated)	Savings
Gas used	Unknown	5400kWh/year (27kWh/m^2)	
Wood fuel used	Unknown	1400kWh/year (7kWh/m^2)	
Total electricity used	Unknown	4980kWh/year (24.9kWh/m^2)	
Total house CO$_2$ emissions	Unknown	1600kg/year (8kg/m^2)	
Space and water heating	478kWh/m^2/year	38.6kWh/m^2/year	90%

Costs

The total cost of the refurbishment was in the region of 168,000 Euros without tax, or 840 Euros per square metre. This cost covered all the works, including extensive internal works, such as the demolition of all partitions and reconstruction of a new layout.

Planning Issues

Since all of the works were internal, no application for planning permission was needed.

Conclusions

The view from the road is unchanged and the new interior provides a new richness of light, space, views out over the garden and a high quality of finishes. The refurbishment costs were considered to be very reasonable, making this a good example of how to achieve good energy performance in large house in a protected urban environment.

Figure 4.9 View from the garden after refurbishment

Source: Myriam Jacquet and Gregoire Magnien

Case Study 4.3 UK: Hereford

- Detached family house built in 1869.
- Radical refurbishment to near *PassivHaus* standards.
- External insulation, triple-glazed windows.
- Whole-house mechanical ventilation and heat recovery.
- Attention to airtightness – achieving 0.97 air changes per hour (ACH).
- Heating demand 18kWh/m^2/year; whole-home CO_2 emissions of 22kg/m^2/year (15kg/m^2 with solar thermal, yet to be fitted).
- Work carried out with occupants in place.

Figure 4.11 House after refurbishment (porch and plaque still to be reinstated)

Source: Andy Simmonds

Figure 4.10 House before refurbishment

Source: Andy Simmonds

Description

Grove Cottage is a two-storey detached house in Hereford, built in 1869 for a railway carriage inspector. It faces east–west, with solid brick walls and a slate roof. Originally it was a 90m^2 two-bedroom house with a 30m^2 basement. The co-owner is an architect and carried out the renovation and extension as a home for his family. The project aimed to reduce CO_2 emissions by 85 per cent compared with the average UK housing stock.

Refurbishment Process

The owner had lived in the house since 2005 with minor changes and found it very difficult to achieve and maintain acceptable temperatures in such a poorly insulated and draughty house: the heat and power consumption was responsible for emitting around 13 tonnes of CO_2 per year. A full refurbishment was undertaken, aiming to achieve *PassivHaus* standards as far as possible, and design work started in 2007. At the same time, a 40m^2 extension was added (comprising a new hall, kitchen and bedroom) and the loft space was converted into usable space. The refurbishment work was carried out with the residents remaining in the house virtually the whole time, and was completed in 2008.

This transformation of an old, virtually uninsulated and draughty house required a radical approach and rigorous detailed design work, combined with a high quality of building work. The initial design was based on low-energy 'rules of thumb' and the design expertise that the architect had acquired over three years of experience gained at the AECB's (The Sustainable Building Association's) CarbonLite Programme. Additional advice was provided by energy consultant and service engineer colleagues. This enabled planning permission to be sought and obtained in 2008, after which the design was modelled for comfort levels and energy performance using the *PassivHaus* Planning Package (PHPP), with some modelling of thermal bridges in difficult areas of construction using 'Therm'. This modelling informed and refined the detailed design of the refurbishment as well as the extension, and showed that very little of the original design had to be modified in order to reach the very stringent targets.

Sustainability Improvements

The new rear extension enabled several high-level south-facing windows to be added to capture solar gains effectively during the heating season and to maximize daylighting year round.

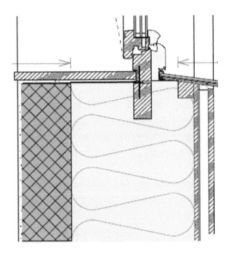

Figure 4.13 Window sill detail

Source: Andy Simmonds

Figure 4.12 Kitchen after refurbishment showing south daylighting

Source: Andy Simmonds

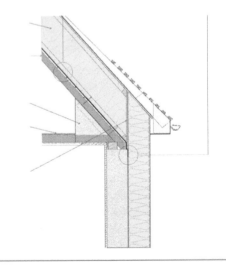

Figure 4.14 Wall and roof junction detail

Source: Andy Simmonds

Table 4.6 Sustainability improvements in Case Study 4.3

Walls		
West and south façades	• External 250mm Permarock expanded 'Neopor type' polystryrene insulated render system, adhesively bonded and mechanically fixed to the masonry walls.	U value: 0.12W/m²K, including the mechanical fixings
East façade	• External 350mm Larsen Truss system, based on AECB Gold Guidance details: timber trusses fixed to masonry and filled with semi-rigid glass mineral wool, sheathed with a timber board and low vapour-resistance membrane, and covered with locally sourced timber cladding on counter battens.	U value: 0.12W/m²K, including the timber frame elements within the insulation layer
Floors (three systems)	• Floating floor with 100mm rigid polyurethane insulation on existing concrete slab.	U value: 0.21W/m²K
	• New insulated concrete raft with no foundations/downstands with 250mm high-density load-bearing expanded polystyrene insulation directly below.	U value: 0.13W/m²K
	• Existing suspended timber floor/basement ceiling: floorboards retained and joists insulated between and below with 225mm sheep's wool and plasterboard lining.	U value: 0.17W/m²K
Roof (two systems)		
New A frame system over existing cottage roof	• Existing 75mm x 55mm rafters retained in place: new layer of sawn timber planks with air-vapour membrane over, then new 400mm deep I beams fixed to old roof at 920mm centres. I beam rafters fully filled with Crown Rafter Roll 32 mineral fibre batts.	U value: 0.09W/m²K
New flat roofs and 30° roof to extension	• 400 mm insulation between I beam rafters, ventilation battens and timber deck, with heat-bonded recycled roofing membrane covered with meadow flower turf.	U value: 0.09W/m²K
Windows	High-performance, triple-glazed inward-opening timber windows and fully glazed doors with insulated frames, faced externally in painted aluminium (based on EDITION range from Internorm).	Full window U values range from 0.75 to 1W/m²K depending on window size
External doors	Fully glazed to the same specification as above.	
Other	• The 25mm–40mm gap to the gable end of the next house (north side) was filled with expanding polyurethane.	
	• Cold bridging was eliminated in the new parts and reduced dramatically in existing parts by additional insulation and careful detailing and construction.	
Airtightness	Simple clear design strategy with special care given to sealing at all junctions in order to reduce air leakage so that a continuous airtightness plane was created throughout the entire house.	Blower door tested air change rate: 1ACH, or 0.97m³/m²/hour
Ventilation	*PassivHaus*-certified whole-house mechanical ventilation system with heat recovery (MVHR). Air extracted from kitchen, bathrooms, clothes drying cupboard and upper hallway and passed through a large counter-flow heat exchanger. Fresh preheated air fed into the living rooms and bedrooms (Paul Thermos 200). Hot weather bypass enables automatic night-time cooling.	Heat recovery: 92%, fan power 53W
Heating and hot water	The existing radiators were retained. The existing boiler was replaced with a small gas boiler, including a low-energy circulating pump, a new mains pressure 300 litre twin-coil, solar-ready hot water cylinder with 50mm polyurethane (PU) insulation, and an additional 100mm mineral fibre lagging added on site. All hot and cold pipes throughout the house were insulated (not radiator system pipes).	
Renewable energy sources	Preparations were made to install solar hot water panels at a later date.	Would reduce house CO_2 emissions from 22 to 15kg CO_2/m²/year

Figure 4.15 New roof truss ready to receive insulation

Source: Andy Simmonds

Comfort and Overheating

Predicted overheating was low due to the external insulation, maintaining the internal thermal mass and the relatively small area of south-facing windows.

The experience of living in the house has supported this prediction, with no overheating occurring. Both MVHR and natural ventilation options for summertime cooling work well when tried. Temperature data logs and personal experience show that comfort conditions (temperature and air quality) were successfully maintained throughout all parts of the house, including the cold winter of 2009/2010.

The monitoring period for temperature and humidity ran from 4 October 2009 to 14 May 2010, with the loggers capturing readings every hour. The minimum temperatures were –6.0°C outdoors and 17.8°C indoors. The maximum indoor relative humidity, averaged over the three rooms monitored, was 75.6 per cent, and the maximum temperatures were 20.2°C outdoors and 23.7°C indoors.

Change in Energy Use and CO_2 Emissions

Table 4.7 Change in energy use before and after refurbishment in Case Study 4.3

	Before refurbishment (measured 2005)	After refurbishment (measured 2009/2010)	Savings
Gas used (gas per square metre)	28,000kWh/year (289kWh/m²)	6206 kWh/year (45kWh/m²)	78% reduction (84% reduction)
Electricity used	13,800kWh/year	4365kWh/year	68% reduction
CO_2 emissions	11,256kg	3046kg	73% reduction
Normalized gas use based on ten-year average temperatures	27,392kWh/year (282kWh/m²)	5136kWh/year (37kWh/m²)	81% reduction (87% reduction)

Notes: The floor area of the house was increased from 97m² before refurbishment to 137m². The period of measured gas use covers the cold winter of 2009/2010, an unusually cold year (2233 degree days compared with the ten-year average of 1848 for the region). Electricity use was measured here from June to December 2009, and January to May 2010. An electric tumble dryer was in use during the earlier period and was later replaced by the successful MVHR drying cupboard.

Costs

The total cost of the extension plus the refurbishment was in the region of UK£125,000; but the project also received elements of commercial support from a number of suppliers (Permarock Ltd, Knauf Insulation, Vencil Resil, Second Nature, Internorm Windows UK, Green Building Store, East Midlands Insulation, City Roofs, Keim Paints, Cemex Concrete Products). Without a detailed costing analysis to help differentiate between new-build and refurbishment costs, and costs above and beyond 'standard refurbishment' costs (e.g. rewiring, wall repairs, etc.), the costs quoted here should be treated with care and will not be representative for different houses!

Planning Issues

There was little change in the front façade, although the building stands out by 250mm from the street building line (which varies between groups of properties) and the roof line was raised (the new rear extension did involve a material change). After negotiation with the planners concerning the potential visual impact affecting neighbours of the rear extension, planning permission was quickly granted and the planners and council executive were very positive about the proposed sustainable refurbishment.

Conclusions

Occupants have found the refurbished house consistently warm in winter and cool in summer; indoor spaces are sunny with high levels of daylight during most types of weather. The MVHR system maintains a very fresh internal atmosphere throughout the year, with no draughts or intrusive noise. No overheating has been experienced. The solar water heating system will be installed as soon as it can be afforded. The only aspect which currently requires change or improvement is a readjustment of the remote boiler control panel/temperature sensor in order to allow easy control of the hot water tank's water temperature.

Figure 4.16 Thermal imaging demonstrates the value of insulation compared with uninsulated houses next door

Source: Andy Simmonds

Case Study 4.4 UK: Culford Road, London

- Four-storey Victorian terraced house, with loft conversion.
- Followed *PassivHaus* principles as far as possible.
- Double- and triple-glazed windows.
- Minimal space heating system, with no heat emitters except on the lower ground floor and first floor bathroom towel rail.
- Whole-house ventilation system with heat recovery.

Figure 4.17 The front of the house has changed little since refurbishment

Source: James Brittain

Description

This terraced house in north London was built circa 1830 and is in a conservation area. The front façade could not be changed in appearance; but it was possible to extend the rear elevation and construct a new study/bedroom in the attic, not visible from the front. The house required a complete renovation, including re-roofing, new wiring, new kitchen and bathrooms, and total redecoration.

Refurbishment Process

The house was bought by the owners in 2007 and the refurbishment work was completed in 2010. The house began as a three-storey dwelling with a gross internal area of 106m², ending up as four storeys and an internal area of 121m², having demolished about 8m² of old extensions, added approximately 33m² of new extensions at the rear and on the roof, and lost 10m² of floor area to internal insulation. *PassivHaus* principles were adopted as far as possible.

Sustainability Improvements

The floor joists on the front were supported on new steel joists between the party walls to avoid cold bridging and to preserve joists. Most cold bridges on the rear elevation were eliminated by adopting AECB Silver Standard details around windows and using foam glass blocks at critical slab-to-wall junctions. Existing chimneys were capped at the top and in each room to stop water ingress and heat loss.

Comfort and Overheating

The house has been occupied since 12 January 2010. Internal temperatures are unusually stable. The summer was not especially hot; but solar gain to the top floor study was sufficiently strong to justify installation of an external motorized blind to control overheating and glare. A reset housing and

Table 4.8 Sustainability improvements in Case Study 4.4

Walls	• Existing solid front walls, 130mm internal glass wool insulation within a double sandwich of oriented strand board (OSB) with insulated studs, with 30mm ventilation gap.	U value: 0.2W/m²K
	• New rear walls, cavity fully filled with 200mm of glass wool insulation, medium density thermal blockwork and low-conductivity boronite cavity wall ties. External skin of recovered bricks.	U value: 0.15W/m²K
	• Party walls insulated (for acoustic reasons and to retain heat if the neighbours turned their heating off), 60mm of insulation fitted between metal studwork.	
Floors	New solid ground floor with 100mm extruded polystyrene insulation.	U value: 0.18W/m²K
Roof	180mm Celotex rigid insulation.	U value: 0.15W/m²K
Windows	• Front sash windows: 4mm argon-filled double glazing with low e; each sash fitted with four draught seals.	U value: 1.8W/m²K
	• New rear windows: inward-opening tilt and turn, argon filled, triple glazed with low e coatings and double seals.	U value: 0.9W/m²K
External doors	50mm insulation.	U value: 1.0W/m²K
Lighting	Fluorescent tubes and compact fluorescents: LED down-lighters, 400W total (3W/m²).	
Other sustainability features	• Sedum roof on rear extension. • Rainwater collection in 1000 litre tank in the garden. • Low-flush toilets.	
Airtightness	Extensive sealing of all gaps and holes, tested with four blower door pressure tests.	Blower door tested air change rate: 1.3ACH (after plastering)
Ventilation	Mechanical ventilation with heat recovery, 'Itho HRU Eco 4', with three speed fans giving 0.25 to 0.75ACH.	Heat recovery: 91%, fan power 25W–180W for supply and extract
Heating	Under-floor heating on lower ground floor, towel rails in the bathrooms (no other heat emitters) from 12kW condensing gas boiler.	
Renewable energy sources	1.32kW peak photovoltaic array.	

wiring had been allowed for in anticipation that a blind would be necessary, so it was straightforward to install. During the unusually cold spell in late November/early December 2010, temperatures in the unheated spaces dropped to 18.5°C. This entails wearing warm indoor clothes, but is perfectly acceptable. Comfort will be enhanced in the upper-ground living room after lined curtains are fitted to the windows at the rear, as this will mitigate radiative losses to a clear night sky.

Figure 4.18 Rear façade after reconstruction

Source: James Brittain

Change in Energy Use and CO₂ Emissions

The annual space heating demand of the house is calculated by the *PassivHaus* Planning Package (PHPP) to be 21kWh/m², which compares with 15kWh/m² for the new-build *PassivHaus* standard. Modelled consumption is presented in Table 4.9.

The house will be comprehensively monitored to provide a real measure of performance.

Costs

The cost of the sustainability improvements alone was estimated at UK£60,000. This includes the cost of the photovoltaic array, which was about UK£9000, including the support frame and installation.

Planning Issues

The house is in a conservation area. The nature of the rear extension work, coupled with the level of intervention involved in the project, led to negotiations with the local planning authority lasting almost 12 months, despite no objections being lodged by neighbours. The narrow-gap double glazing on the front façade was used to avoid any change in appearance, and the new roof and photovoltaic array were designed so as not to be seen from the front.

Conclusions

The rating fell between the AECB Silver and Gold Standards for new houses. The careful architectural design work, together with attention to maximizing insulation levels and minimizing air leakage, produced a very attractive daylit low-carbon family house.

Table 4.9 Change in energy use before and after refurbishment in Case Study 4.4

	Before refurbishment	After refurbishment	Savings
Gas used	28,082kWh/year	5985kWh/year	79% (22,097kWh/year)
Electricity used	2592kWh/year	2334kWh/year	10% (258kWh/year)
Total CO₂ emissions	6.54 tonnes CO₂/year	2.15 tonnes CO₂/year	67% (4.4 tonnes CO₂/year)
Cost savings (approximate)		UK£100 heating	UK£1300/year

Note: figures exclude a contribution from the PV array, estimated at 1000kWh/year (0.6 tonnes CO₂/year).

Figure 4.19 View out into the garden

Source: James Brittain

Case Study 4.5 UK: Carshalton Grove, London

- 1870s solid-walled brick house.
- 75 per cent saving in CO_2 emissions achieved.
- Range of wall insulation systems used.
- Ceiling moulding remade.
- Sustainable improvements accounted for only 15 per cent of the total refurbishment cost of UK£85,000.
- Under-floor heating provided in all rooms.
- Solar water heating for domestic hot water using evacuated tube collectors.

Figure 4.20 The appearance of the house from the front has not changed

Source: Russell Smith

Description

This traditional semi-detached, three-bedroomed Victorian house in south London was built in 1870 and was acquired by the current owner in 2005 with the intention of undertaking a comprehensive sustainable refurbishment in order to provide a family house. The house was structurally sound but suffered from problems with damp. The house is built of brick and although not located in a conservation area, the owner did not wish to change the appearance with external insulation. A recent small extension to the rear was constructed in cavity walling.

Refurbishment Process

The house was completely refurbished, starting in 2007, and the loft was converted into a bedroom. Other internal improvements were also made to the rear extension, and ceiling mouldings were remade in the front room. The owner wanted to investigate a number of different insulation systems to explore the costs, ease of use and effectiveness of the finished products. The owner carried out some of the refurbishment works himself and was closely involved in all the works in order to understand the appropriateness of all the systems and products.

Figure 4.21 Internal insulation on the front wall

Source: Simon Burton

Sustainability Improvements

Table 4.10 Sustainability improvements in Case Study 4.5

Walls	Several different internal insulation systems:	U values:
	• Metal frame with 100mm polyisocyanurate and 72mm polyurethane bonded to plasterboard.	0.13W/m²K
	• 100mm wood fibre.	0.33W/m²K
	• 175mm sheep's wool on timber frame with ventilated cavity.	0.21W/m²K
	• 150mm polyisocyanurate on timber battens, with services void.	0.13W/m²K
	• Timber frame with 175mm mineral wool, Isonat or sprayed cellulose fibre.	0.22W/m²K
	• 200mm Isonat in timber stud.	0.20W/m²K
	• 135mm Rockwool in timber stud.	0.31W/m²K
Floors		U values:
	• Some new solid floor with 300mm polystyrene.	0.15W/m²K
	• Existing: 175mm cellulose fibre, supported under timber flooring.	0.14W/m²K
Roof	200mm cellulose fibre in both flat and sloping roofs.	U value: 0.16W/m²K
Windows		U values:
	• Mostly double glazed with argon fill and low e coatings in sash windows.	1.4W/m²K
	• Triple-glazed window at rear.	0.8W/m²K
	• Double-glazed Velux roof lights with moveable external shades.	1.3W/m²K
External doors	Rear French doors added (Rationel); draught lobby on front door.	U value: 1.2W/m²K
Other sustainability features	Rainwater collection for toilet flushing.	
Airtightness	Care taken to seal around windows, doors, floors and skirting boards.	Unknown
Ventilation	Natural ventilation with trickle vents above windows and heat recovery room extract fans/ventilators in kitchen and bathroom.	
Heating	Condensing gas boiler with under-floor heating on all floors.	
Renewable energy sources	4m² of evacuated tube solar collectors for hot water.	

Comfort and Overheating

The house is now very comfortable due to the highly insulated walls, floors and roof, and the under-floor heating. No overheating has been experienced, helped by the external shades on the roof lights.

Figure 4.22 Ceiling mouldings were replaced where covered by insulation

Source: Simon Burton

Figure 4.23 New sash windows with double glazing and sealing

Source: Simon Burton

Change in Energy Use and CO₂ Emissions

Table 4.11 Change in energy use before and after refurbishment in Case Study 4.5

	Before refurbishment (based on modelled data)	After refurbishment (based on modelled data)	Savings
Gas used	305kWh/year	34kWh/year	89% (261kWh/year)
Electricity used	93kWh/year	56kWh/year	40% (37kWh/year)
CO_2 emissions	7.8 tonnes CO_2/year	2.4 tonnes CO_2/year	75% (5.4 tonnes CO_2/year)
Cost savings (approximate)			UK£1278/year

Costs

The full refurbishment costs for the works amounted to UK£85,000 of which the sustainability works ware estimated by the owner to be around 15 per cent (UK£12,750), mostly made up by the upgraded double glazing, the under-floor insulation and the solar domestic hot water installation. The owner believes that the low additional costs are due to taking every opportunity to install insulation and sustainable components when other works were being carried out.

Planning Issues

The only issue on which planning permission was required was the loft conversion and extension.

Conclusions

The renovation has produced a very attractive house with increased floor area, despite the losses from the internal insulation used on all the external walls. New ceiling mouldings were added on the front wall in the living room where the original ones were covered by the insulation, and the internal insulation around the windows was splayed back to improve daylighting. The wood fibre insulation proved very easy to use and was robust enough for fixing cupboards, while other materials were more difficult to fix into place. An early form of Aerogel insulation was also tried out in a small area.

The house is now also very quiet and provides even temperatures, with no draughts, due to the under-floor heating, the removal of cold surfaces and the elimination of unwanted air movements. The house is now open to the public as a demonstration of sustainable refurbishment, and the owner now runs a small company which provides a range of services to people wishing to carry out similar works to their houses (contact Russell Smith of Parity Projects at russell@parityprojects.com).

Case Study 4.6 UK: Chester Road, London

- Victorian semi-detached house completely renovated.
- Original windows double glazed and draught-stripped; new windows double glazed, argon filled and low e coated.
- Walls internally insulated with 100mm or 50mm fibreboard.
- Extensive insulation of roof and floor.
- Draught lobby added in hall.
- Solar thermal system.
- Low-energy lighting throughout and AAA equipment.
- Greywater and rainwater recycling.
- Wood-burning stove in the living room.
- Solar PV system added.

Figure 4.24 The front façade is basically unchanged since refurbishment

Source: Jeffery Pike

Description

This late Victorian semi-detached house in a north London conservation area was bought by the owner in 2004 and required a total refit. The roof was leaking, the heating was barely functioning and the house was freezing cold, with a lot of condensation. There was no shower, the kitchen was on its last legs and the house needed redecorating throughout. The owner wished to demonstrate that these types of houses do not have to be pulled down; it can, in fact, be cheaper fiscally and energy-wise to make a Victorian house energy efficient.

Refurbishment Process

The owner took the decision to carry out comprehensive energy improvements while the whole house was being renovated and to address all sustainability issues. Since the house was in a conservation area, the front façade was to remain virtually unchanged, and internally the owner wished to retain all period features, particularly the ceiling mouldings. Where the internal wall insulation covered the mouldings, they were remade by a specialist, and the existing sash window frames were retained and renovated with new double-glazed

Figure 4.25 Floor and wall insulation in progress

Source: Jeffery Pike

units and draught-stripping. The main refurbishment works were carried out in 2006.

The owner now runs an eco-refurbishment consultancy business, using her house as a show house (contact Sarah Harrison of Eco-Refurbishment at www.eco-refurbishment.co.uk).

Sustainability Improvements

Table 4.12 Sustainability improvements in Case Study 4.6

Walls	All external walls internally insulated with 100mm or 50mm fibreboard	U value: 0.31–0.55W/m²K
Floors	• Original wooden flooring on the living-room ground floor removed (and reused in the attic) and 170mm of under-floor insulation added (150mm mineral wool between joists and 20mm wood fibre pinned to underneath of joists).	U value: 0.23W/m²K
	• Solid floor in kitchen extension insulated with 100mm closed cell polystyrene.	U value: 0.11W/m²K
	• Original Victorian tiled hall floor not insulated in order to preserve it.	
Roof	• Whole of main roof insulated with 150mm mineral wool between rafters and two layers (75mm) of expanded wood fibreboards over joists.	U value: 0.17W/m²K
	• Extension roof insulated with 250mm mineral wool.	U value: 0.16W/m²K
Windows	• Original sash windows double glazed using argon-filled low e glass and draught-stripped.	Centre pane U values: 2.0W/m²K
	• New rear windows double glazed, argon filled and low e coating.	1.2W/m²K
External doors	• Original front door retained but draught-stripped.	U value: 2.2W/m²K
	• High-performance new back door.	
Other sustainability features	• Low-energy lighting throughout. • Sunpipe to allow natural light into the hallway. • AAA washing machine, dishwasher and fridge freezer. • Greywater and rainwater recycling for garden use. • The house was plumbed for water efficiency using flow regulators, aerating taps and showers, and ultra-low water usage WCs.	
Airtightness	Whole house draught-stripped and a draught lobby installed in the hall.	Blower door tested air change rate: 5.6ACH
Ventilation	Heat recovery extract fans in bathroom, drying cupboard and shower room.	Heat recovery fan up to 80% heat recovery; fan normal operating 2W
Heating	• Condensing gas boiler. • Wood-burning stove in the living room (installed in 2008).	
Renewable energy sources	• 2m² of flat-plate solar-collector system for domestic hot water, with PV-powered pump.	75% of annual DHW
	• 10m² of PV panels connected to National Grid (installed in 2010).	Estimated annual generation 1500kWh

Figure 4.26 Rainwater 'wall' in the garden

Source: Jeffery Pike

Comfort and Overheating

As a result of the heat-conserving measures and the wood-burning stove in the living room (which is fed with waste wood), there is rarely a need to switch on the central heating, which is then only used for an hour or two in the morning and evening during winter. The occupants can sit by the windows in winter without ever feeling cold. There has been no experience of overheating in summer.

Figure 4.27 Ceiling mouldings remade to match the original where necessary

Source: Jeffery Pike

Change in Energy Use and CO_2 Emissions

Table 4.13 Change in energy use before and after refurbishment in Case Study 4.6

	Before refurbishment	After refurbishment (monitored)	Savings
Gas used	Unknown	5500kWh/year	
Electricity used	Unknown	1400kWh/year	
CO_2 emissions (calculated)	9.8 tonnes CO_2/year	1.8 tonnes CO_2/year	82% (8 tonnes CO_2/year)

Notes: Energy uses have only been monitored after refurbishment; CO_2 emissions are calculated. These do not take account of the PV panels, which are newly installed.

Costs

The refurbishment to this high environmentally friendly specification cost approximately UK£17,000 more than the standard practice alternatives. The owner believes that much of the green technology installed has fallen in price during the intervening years since the house was completed, and that many homeowners would happily spend what was allocated to green improvements on a new kitchen. Not only has value been added to the property (which will become much more apparent in the coming years), but fuel bills have been dramatically cut.

Planning Issues

Since the house was in a conservation area, the appearance of the front façade could not be changed – hence, the refurbishment of the existing sash windows and the internal wall insulation. The solar thermal panels were placed on the south-east facing back-extension roof, and were given planning permission in 2006 as they were not visible from the front.

The wood-burning stove meets the Clean Air Regulations.

The PV panels were installed in 2010; since they are not visible from the front, they are deemed to fall within the terms of permitted development and so did not require planning consent.

Conclusions

From the outside, this three-bedroom home looks like a typical north London red-brick Victorian semi. Step inside and, to the untrained eye, there is little evidence that it has been renovated using green energy technology. People are surprised that it looks and feels like a nice, but normal, Victorian house, while achieving 80 per cent energy-efficiency savings through green refurbishment. They expect a modern look, and are surprised that you can make a home energy efficient without giving up the period features. Most of all, people are surprised at how cheaply the carbon saving goals have been reached.

The owner regrets that she did not divert the rainwater from the back roof into a tank in the extension loft and use this to flush the WCs on the first and ground floors. This would have been a low-cost thing to do as part of the refurbishment and would have increased water savings to approximately 50 per cent. It would now be a disruptive and relatively expensive thing to do.

Figure 4.28 Photovoltaic panels, also on top of the dormer window

Source: Jeffery Pike

Case Study 4.7 UK: Bristol

- Large solid-walled house built in 1934.
- External insulation to all external walls.
- Ground floor and roof insulation.
- Reuse of materials.
- Reduction in water use.
- Solar thermal hot water and PV panels installed.
- Low-energy lighting.
- Garden used to grow vegetables.

Description

This detached house in Bristol in south-west England was bought by architect owners in 2004 and renovated to provide a home for their family and an office for their architectural practice. The house had 225mm solid brick walls, a wooden roof, leaky metal-framed single-glazed windows, and was completely uninsulated.

Refurbishment Process

The owners wanted to use the project as a laboratory for some of their more challenging sustainability ideas and also wanted to collect as much data as possible for the house once it had been finished to inform the eco-refurbishment debate. Renovation was carried out during 2005/2006, including the conversion of the roof space into the office and construction of a green oak canopy over a deck in the garden.

Figure 4.29 House before renovation

Source: Lucy Pedler, archipeleco architects

Figure 4.30 House after renovation

Source: Lucy Pedler, archipeleco architects

Figure 4.31 Applying the external insulation

Source: Lucy Pedler, archipeleco architects

Sustainability Improvements

Table 4.14 Sustainability improvements in Case Study 4.7

Walls	• External insulation on all four façades; 80mm polyisocyanurate with three-coat render. • Tile-hung section (and chimney breast) internally lined with 20mm foam-backed plasterboard.	U value: 0.22W/m²K
Floors	150mm cellulose insulation in between joists; breather membrane holding insulation in.	U value: 0.2W/m²K
Roof	100mm mineral wool underlay insulation between rafters and 70mm overlay board on top of rafters.	U value: 0.22W/m²K
Windows and patio doors	Double glazed, low e coated, argon filled in timber frames (Rationel windows).	Full window U value: approximately 1.33W/m²K
External doors	Front door, original with double layer of draught-proofing and foam-backed plasterboard within panels of door on inside.	
Lighting	Mixture of LEDs, compact fluorescents and halogen/incandescent lights (being phased out).	
Other sustainability features	• Use of reclaimed wood for floors. • Water-saving measures. • Clay and organic paints, linoleum and slate used.	
Airtightness	• Ground-floor, airtight waterproof membrane lapped up sides of all external walls. • Roof: airtight waterproof membrane.	Blower door tested 8m³/m²/hour
Ventilation	Natural trickle vents in all windows; extract fan in new shower room.	
Heating	• Gas condensing boiler for space heating and winter domestic hot water. • Solar thermal system.	
Renewable energy sources	• 4m² solar thermal panels for domestic hot water. • 1.225kW peak PV panels. • 50% DHW. • 25% of total electricity demand of house plus office.	

Figure 4.32 Photovoltaic array on rear roof; solar thermal panels are located on the dormers

Source: Lucy Pedler, archipeleco architects

Comfort and Overheating

All of the occupants agree that the house is a very comfortable place in which to live and there has been no overheating during summer.

Change in Energy Use and CO_2 Emissions

Costs

The total cost of the refurbishment works was more than UK£100,000 and the additional sustainability costs cannot be estimated separately (e.g. new windows were needed anyway). However, the external insulation is known to cost approximately UK£16,000 to £18,000.

Table 4.15 Change in energy use before and after refurbishment in Case Study 4.7

Energy use in the house only	Before refurbishment (typical UK usage for the size of house)	After refurbishment (monitored)	Savings
Gas used	38,400kWh/year	12,148kWh/year	68% (26,252kWh/year)
Electricity used	11,520kWh/year	2799kWh/year	76% (8721kWh/year)
CO_2 emissions	12.3 tonnes CO_2/year	3.5 tonnes CO_2/year	71% (8.8 tonnes CO_2/year)

Planning Issues

The external appearance of the house has not changed even though nearly 100mm of insulation and render have been added. There were no planning implications and no restrictions on the solar and PV panels, though the new dormer windows required approval from the planners.

Conclusions

Occupants have found the refurbished house easy to warm up from cold, and the insulated thermal mass of the external walls retains the heat very effec-

tively. The only aspect which requires change or improvement is the temperature differential between the ground and second floors. During winter, the ground-floor radiators are on full; the first-floor thermostatic radiator valves (TRVs) are set at 2 to 3; and the top floor has no radiators on, but due to the open-galleried staircase and the double layer of roof insulation, the heat rises and is trapped at the top. Engineers have been engaged to look into this. The owners would also like to capture the passive solar heat from the conservatory and redistribute it around the house (engineers are also looking into this).

Figure 4.33 View from the garden

Source: Lucy Pedler, archipeleco architects

Case Study 4.8 UK: Manchester

- Large, solid-walled semi-detached house.
- Sash windows refurbished and double glazed; some windows triple glazed.
- Flank wall externally insulated with 170mm wood fibre.
- Aerogel internal wall insulation chosen for front and rear walls.
- Wood-fired gasifying boiler with large thermal stores.
- Wood-burning stove.
- Solar thermal evacuated tube collectors.
- Low-energy lighting.
- Environmentally friendly products and systems.

Description

This completely uninsulated, solid-walled, semi-detached Victorian house was converted from an individual room house back to a family house in order to accommodate the owners' family. The house was in need of full renovation but had internal ceiling mouldings, which were retained, and rooms in the attic space. The owner worked very hard in choosing the most environmentally friendly insulation materials and aims eventually to make the house zero carbon. Heating is by a gasifying wood furnace.

Figure 4.34 Front façade before renovation

Source: Charlie Baker

Figure 4.35 Front façade after renovation

Source: Charlie Baker

Refurbishment Process

The house was purchased in 2004 and most of the works were completed by 2009. The occupants lived in the house during most of the refurbishment works, which were undertaken sequentially in order to minimize health hazards (e.g. starting with rewiring). Some insulation works and kitchen replacements are still being carried out. The aerogel internal-wall insulation system was chosen for the front and back walls to give the highest insulation within the 40mm thickness of the old plaster, which was removed. The key disadvantage was cost; but the insulation could be fixed directly to the wall with no framing or gluing, so the cost of the material could be offset by the ease of installation. The choice of wood fibre over mineral wool for the external insulation on the gable end was based on vapour permeability, sequestered carbon, the benign properties of wood fibre, and the manufacturer's sustainability efforts through the entire product (e.g. glues, etc.).

Figure 4.36 External insulation in progress on flank wall; detail at window

Source: Charlie Baker

Figure 4.37 Flank wall after insulation

Source: Charlie Baker

Sustainability Improvements

Table 4.16 Sustainability improvements in Case Study 4.8

Walls	• Gable wall, externally insulated and rendered 170mm recycled wood fibre.	U value: 0.2W/m²K
	• Front and back walls, internally insulated (fitted later) 27mm Spacetherm (aerogel system) glued to board.	U value: 0.38W/m²K
Floors	G200mm glass wool under the ground floor (taking down the cellar ceiling).	
Roof	• Flat loft section, 350mm formaldehyde-free glass wool.	U value: 0.1W/m²K
	• The main slope of the roof: rafters built up to accommodate 200mm sheep's wool with plasterboard and skim (no vapour barrier, so fully vapour permeable).	U value: 0.14W/m²K
	• Area under the half gable and dormers (where space was limited): 100mm rigid glass wool batts, with 27mm of Spacetherm on board (vapour impermeable).	U value: 0.15W/m²K
Windows	• New windows: triple glazed with argon fill in Forest Stewardship Council (FSC) hardwood frames with double seals.	New triple-glazed windows, U value: 0.9W/m²K
	• Existing windows: refurbished with low e-coated, double-glazed units or secondary glazing to original stained glass.	
	• All new and existing windows draught-stripped with Neoprene O-profiles routed into the casements.	
External doors	No new doors fitted.	
Lighting	Generally, compact fluorescent lamps, spotlighting with high-intensity discharge (HID) lamps, down-lighting mainly with dimmable, recessed compact fluorescents.	
Other sustainability features	Low-flow spray taps where new.	
Airtightness	Not tested.	
Ventilation	Natural.	
Heating	• 40kW gasifying log burner and large solar thermal collectors both feeding into heat stores providing almost all hot water and some space heating.	
	• Log-burning stove.	
Renewable energy sources	10m² of solar thermal collectors using 80 evacuated tubes in four arrays.	

Comfort and Overheating

The house has proved to be warm in winter and is unlikely to overheat due partially to its aspect and footprint. The highest internal temperature recorded in the summer of 2010 was 26°C.

Change in Energy Use and CO_2 Emissions

Table 4.17 Change in energy use before and after refurbishment in Case Study 4.8

	Before refurbishment (calculated)	After refurbishment (calculated)	Savings
Gas used	87,180kWh/year	1500kWh/year*	
Wood used	–	41,730kWh/year	
Electricity used	6480kWh/year	5640kWh/year	13%
CO_2 emissions	19.6 tonnes CO_2/year	3.6 tonnes CO_2/year	82%

Note: * Gas used after refurbishment for cooking only.

Costs

The total cost of the refurbishment to date was in the region of UK£85,000, including a new roof, complete rewiring, a new burglar/smoke alarm system, two refurbished bathrooms, two fully rebuilt bedrooms and an office, as well as the energy measures. The carbon reduction works, the insulation, the draught-proofing and the heating cost UK£38,500. The works included high-quality fittings and equipment, in general, and the heating system itself (a gasifying wood burner and water storage systems) cost UK£7234, while the solar water system with evacuated tubes cost a further UK£7182. The owner thinks that the purchase price plus the refurbishment works are less than he would have paid for a house this size in good order. Purchasing tree prunings for the boiler is about two-thirds of the cost of buying pellets, and some waste wood has been provided by a local builder.

Planning Issues

The log gasifier required local permission due to emissions on start-up; but planning permission was not required for any other works.

Conclusions

The project has been carried out over a number of years. Since the wood stove uses old wood as available and timber bought in from a local tree surgeon, the moisture content varies and this affects the performance of the stove. Photovoltaic panels could be added in the future to move the house towards carbon neutrality.

Figure 4.38 Evacuated tube solar collectors

Source: Charlie Baker

Case Study 4.9 Portugal: Western Algarve

- Rural 30-year-old villa completely refurbished and enlarged.
- Layout and architecture improved.
- Use of solar gains and solar protection an important aspect of the refurbishment.
- Full external wall insulation.
- All elements insulated to high standards.
- CO_2 emissions reduced by 89 per cent.

Figure 4.41 The north façade before renovation

Source: Jes Mainwaring

Figure 4.39 The south façade before renovation

Source: Jes Mainwaring

Figure 4.42 The north façade after renovation

Source: Jes Mainwaring

Figure 4.40 The south façade after renovation

Source: Jes Mainwaring

Description

The original 30-year-old single-storey villa was built to the very poor energy and comfort performance standards prevailing in Portugal until recent legislation. Constructed from single-leaf 240mm typical hollow clay uninsulated block walls with cement render internally and externally, uninsulated concrete and tile roofs, and aluminium-framed single-glazed windows, the villa suffered the endemic local problems of black mould and dampness in winter from condensation and water penetration. Dampness in spaces on the sunless north side was particularly bad to the point of being potentially unhealthy. During summer, the interior suffered from severe overheating, and the building was in critical need of general refurbishment. Typical of many houses in Southern Europe, it was not well laid out in spatial terms, and architecturally was relatively uncoordinated and random.

Refurbishment Process

The concrete frame and hollow clay-block construction of the existing building is typical of Southern Europe, and has been extensively refurbished and upgraded throughout in support of the energy strategy. The house was also extended by a 55 per cent increase in area to 180m². New internal and external walls are built similarly using 'Tabicesa' system blocks in order to take advantage of their thermal mass.

The old villa had a north–south orientation with wide sliding glazed-door openings into the living room and main bedroom on the south side. The western side had limited openings that gained little from the available sun, while the north was in perpetual shade. Two single-room 'wings' extended northwards from the east–west spine wall, and these were extended to allow the courtyard to be enclosed with a glazed roof to form an entrance atrium as part of the strategy to introduce solar gains to the north part of the villa. The shallow pitched glass roof receives huge amounts of direct sunlight throughout the year at this latitude, and has thermostatically controlled automatic vents to prevent overheating. Internal replanning ensures that every room now has direct access from the atrium, which also promotes warm air drift throughout the villa.

Additionally, new tiled pitched roofs

over the extended northern spaces on either side of the atrium have been configured to give significant south-facing pitches. As well as providing a discreet location for the solar water collectors, these incorporate large roof windows that give direct solar gains and natural daylight from above. Two new full-height glazed-door openings were made in the western external wall, adding considerably to the solar gains delivered to the busy daytime areas, and these were fitted with timber shutters for protection against overheating during hot conditions.

To the south, a new structural timber pergola supports a roof overhang where the existing roof has been extended to form a protected and shaded all-weather outside seating area. This also gives shade from the high summer sun to the glazed sliding doors from the main bedroom to the outside terrace, while allowing solar gains during winter. The unroofed part of the south-facing pergola also carries a tensile cable structure to enable newly planted vines to provide bioclimatic shading and Muscatel grapes. This shades the large glazed opening into the living room in summer, but permits full solar gains in winter when the vine leaves have been shed and the grapes have been eaten.

Figure 4.43 Morning sun in autumn providing solar gain to the north side

Source: Jes Mainwaring

Sustainability Improvements

Table 4.18 Sustainability improvements in Case Study 4.9

Walls	External insulation with 60mm expanded polystyrene and thin-coat through-coloured acrylic render. Extends 700mm below ground.	U value: 0.4W/m²K
Floors	New tiled concrete floor slabs, 200mm Leca expanded clay insulated screed.	U value: 0.3W/m²K
Roof	• New roofs 110mm expanded polystyrene timber sandwich panels on timber rafters. • Existing roofs stripped back to concrete structure and insulated with 110mm expanded polystyrene rigid-insulation panels.	U value: 0.32W/m²K
Windows and external doors	New larch double-sealed precision window and glazed door frames, argon filled, low e, double glazed (laminated for security).	U value: 1.2W/m²K
Lighting	All internal and external lighting by either LEDs or low-energy lamps.	
Other sustainability features	• Extensive use of sustainable natural materials. • Bioclimatic shading techniques (rainfall pattern in this climate means that domestic rainwater collection is not viable).	
Airtightness	Precision-engineered double-sealed timber window and external door system eliminates draughts.	
Ventilation	• Natural, with trickle vents over windows. • Presence detection-operated extract fans.	
Heating	Propane gas boiler for space heating and solar domestic hot water backup.	
Renewable energy sources	• 4m² of solar thermal panels. • 8kW wood-burning fire cassette with fan-assisted warm air convection.	

Comfort and Overheating

The comfort levels within the villa are beyond comparison with those previously experienced, where damp, cold conditions prevailed during winter and comfort was difficult to achieve beyond a single room where a log fire provided the only realistic heat source. The wet radiator system that had been retrofitted was hopelessly undersized and could not cope with the heat losses through the building's uninsulated fabric, leaving each space chilled by cold wall surfaces. During summer, overheating was a constant problem, with internal temperatures many degrees hotter by evening than the outside air temperature.

The house now maintains comfortable, consistent and affordable internal temperatures throughout the heating season; even during the hottest recorded summer of 2010, overheating was not a problem as the interior remained comfortably cooler than outside.

Change in Energy Use and CO$_2$ Emissions

Table 4.19 Change in energy use before and after refurbishment in Case Study 4.9

	Before refurbishment (calculated)	After refurbishment (calculated)	Savings
Gas (propane) used	50,407kWh/year	12,829kWh/year	75% (37,578kWh/year)
Electricity used	Unknown	Unknown	
CO$_2$ emissions	2.97 tonnes CO$_2$/year	0.316 tonnes CO$_2$/year	89% (2.65 tonnes CO$_2$/year)

Costs

The works cost significantly less than demolition and rebuilding, both in monetary and carbon terms, and have extended the life of the property to equal that of a new building.

By design, the energy performance and sustainability aspects form an integrated element of the whole project, and costs are difficult to isolate. The overall total project cost of about 170,000 Euros includes the increase in area by 65m^2 to approximately 180m^2, extensive energy measures and general architectural improvements. Approximately 15 to 20 per cent of the project cost was allocated to the refurbishment and remodelling of the existing building, with high standards of quality in the specification of materials, finishes, fixtures and fittings; but these architectural improvements did not contribute to the improvements in energy performance. This leaves between 135,000 and 140,000 Euros for improvements relating to enlargement, energy and sustainability. For comparative purposes, the cost of demolishing and replacing the existing villa with a comparable new-build villa would probably be in the order of 250,000 Euros.

Planning Issues

The villa is set in a rustic landscape which is heavily protected by preservation and environmental controls, and new construction is not permitted apart from in very exceptional circumstances. These conditions required that the project had to be approved by a number of environmental authorities in Lisbon, as well as the local authority. The application included verification of the reduction in carbon emissions and a considerable improvement to the architecture. The required consents took about seven months, but were eventually forthcoming.

Figure 4.44 Inside the northern atrium

Source: Jes Mainwaring

Conclusions

The refurbishment significantly improves the architecture, usability and comfort of the villa by selectively increasing its size in a way that not only improves its internal layout, but also integrates passive design strategies. The planned solar-shading devices for the atrium and roof windows are not yet installed, but would make a useful contribution to keeping the interior cooler in summer.

Architect and owner: Jes Mainwaring, AADipl RIBA.

Energy engineer: Colin Reid, BSc Hons, C.Eng, MCIBSE, Eur. Ing.

Case Study 4.10 US: Oakland, California

- Small American house built in 1915 and renovated in 2008.
- Zero-energy goal using solar water heating and a photovoltaic array.
- 66 per cent reduction in heating compared to standard refurbishment.
- Rainwater and greywater capture and use.
- Highest rating on the Leadership in Energy and Environmental Design (LEED) Homes Green Building Rating System (LEED Platinum).
- Highest-rated LEED for Homes renovation.
- Owner's home office is a 'regenerative lifepod' in the garden.

Description

This 140m² single-storey home was built in 1915 as an artisan's house and was renovated to be a zero-energy house in 2008 by David Gottfried, the founder of the US GBC (Green Building Council) and World GBC. The house is located in the US in Oakland, California, across the bay from San Francisco, close to Berkeley, in Rockridge, a walkable neighbourhood with all facilities provided locally. The house had not been touched in 60 years, so it required a full renovation of its kitchen, bathroom, services, and just about everything else (it still had many of its original Craftsman wood details).

Refurbishment Process

The house was bought in 2007 and, being much smaller than the owners' previous house, needed to be converted into a more space-efficient house by moving and removing walls, changing spaces, opening up the attic and converting the basement into a family room. A large design and construction team was employed to carry out the work. Although the zero-energy goal was important, the renovation included many other aspects of sustainability, including the reuse of old and ecological materials, water use, planting and local supply.

Figure 4.45 The house before restoration

Source: David Gottfried

Figure 4.46 The house after restoration

Source: David Gottfried

Sustainability Improvements

Table 4.20 Sustainability improvements in Case Study 4.10

Walls	90mm borate cellulose insulation fill to external walls.	U value: 0.44W/m²K
Floors	R19 formaldehyde-free batt insulation.	U value: 0.30W/m²K
Roof	• Gable end walls sprayed with polyurethane foam. • Ceiling insulation: 120mm polyurethane foam.	U value: 0.19W/m²K
Windows	New double-glazed Marvin Clad windows with low e and argon fill.	U value: 2.0W/m²K
External doors	New double-glazed Marvin Clad doors with low e and argon fill.	
Other	• All appliances Energy Star, low water and quiet. • Low water-use toilets, showers and taps. • Reused materials, including timber and doors. • Efficient lighting with some LEDs.	
Airtightness	Unknown.	
Ventilation	Bath and kitchen exhaust fans. No cooling system.	
Heating	Gas-fired low-temperature system, incorporating solar thermal with 500 litre storage radiant heating with no fanned air.	
Renewable energy sources	Solar water heating. Photovoltaic panels: 16 panels peak output 2.72kW, as well as 8 panels on the home office.	70% of DHW; 10% space heating

Comfort and Overheating

The house now stays warm in winter and cool in summer. Many houses in the area use cooling during the summer; but no cooling system has been installed in this house and active cooling has not been necessary.

Change in Energy Use and CO_2 Emissions

Table 4.21 Change in energy use before and after refurbishment in Case Study 4.10

	Standard house	After refurbishment	Savings
Gas used for heating	15,080kWh/year	5064kWh/year	66% (10,016kWh/year)
CO_2 emissions	2.8kg CO_2/year	0.9kg CO_2/year	66%

Figure 4.47 Solar thermal and photovoltaic panels

Source: David Gottfried

Figure 4.48 Original Craftsman wood details have been retained

Source: Michael Dambrosia

Costs

The total cost of the refurbishment was in the region of US$400,000 (no separate assessment of the sustainability aspects was made).

Planning Issues

Several of the changes required planning permission.

Conclusions

The owner is unclear how much the sustainability improvements increased the value of the house; but the overall refurbishment definitely did. The green rehab was very successful, but did not result in a total zero-energy building. During the winter months, the building still needs more energy than it generates. There has been significant publicity for the home, which has won several local green building awards, including articles in local and national publications.

Case Study 4.11 Germany: Cologne

- 1920s terraced house.
- Comprehensive refurbishment and remodelling.
- 160mm of external insulation.
- 74 per cent primary energy saving.
- Solar water heating.
- Whole-house mechanical ventilation with heat recovery.

Description

This three-storey terraced house in a Cologne side street was built in 1929 and refurbished by the new owners, a couple with a child. The 162m² house, typical of terraced housing in Cologne, was originally cold due to solid walls and draughts, and the refurbishment and remodelling have produced an ecologically and aesthetically pleasing house that satisfies current environmental standards.

Figure 4.49 Front façade before refurbishment

Source: DENA

Figure 4.50 Front façade after refurbishment

Source: DENA

Refurbishment Process

Extensive works were carried out both inside and outside the house. The transition from living area to garden was given a new open, flowing design, with a generous patio and panoramic windows. Thick external insulation was combined with the insulation of all other external surfaces and modern and environmentally friendly services.

Sustainability Improvements

Table 4.22 Sustainability improvements in Case Study 4.11

Walls	160mm external insulation with render finish.
Floors	80mm insulation under ground floor in the basement.
Roof	240mm insulation in the loft.
Windows	Double glazed with insulating glass.
External doors	Patio doors as windows.
Other sustainability features	Unknown.
Airtightness	Sealing and draught-stripping.
Ventilation	Whole-house ventilation system with heat recovery.
Heating	Gas boiler and solar panels.
Renewable energy sources	4.6m² of solar thermal panels for space and hot water heating.

Comfort

The family is happy that cold walls and draughts are a thing of the past and that they now have a good quality of life.

Figure 4.51 Garden façade after renovation

Source: DENA

Figure 4.52 Front door

Source: DENA

Change in Energy Use and CO$_2$ Emissions

Table 4.23 Change in energy use before and after refurbishment in Case Study 4.11

	Before refurbishment (calculated)	After refurbishment (calculated)	Savings
Total primary energy used	33,534kWh/year	9234kWh/year	72% (24,300kWh/year)
CO$_2$ emissions			74% (11 tonnes CO$_2$/year)

Costs

Not available.

Planning Issues

No planning issues.

Conclusions

The house has attracted attention beyond the close neighbourhood and has spread the message that energy efficiency belongs to refurbishment as a matter of course. The house was included in the Deutsche Energie-Agentur *Efficient Homes* project in 2003.

Case Study 4.12 Germany: Oldenburg

- 1860s traditional gabled house.
- Complete refurbishment.
- 95 per cent primary energy savings.
- Vacuum insulation used in the basement.
- *PassivHaus* standard windows.
- Wood pellet boiler supplies heating.
- Large evacuated tube solar-thermal system.

Figure 4.54 The house after refurbishment

Source: DENA

Refurbishment Process

The owners decided to carry out a comprehensive energy refurbishment, which did not affect the appearance of the house, at the same time as other major improvements. External wall insulation was combined with *PassivHaus* standard windows, and a large solar heating system and a wood pellet boiler supply the heating.

Figure 4.53 The house before refurbishment

Source: DENA

Description

Oldenburg is a small city in northern Germany. The new owners of this large gabled house, typical of Oldenburg, realized that a full refurbishment was necessary to make the house habitable and that they needed more space. This 250m² detached house was built in 1869.

Figure 4.55 The evacuated tube solar collectors

Source: DENA

Sustainability Improvements

Table 4.24 Sustainability improvements in Case Study 4.12

Walls	<ul><li>160mm external insulation on external walls.</li><li>30mm vacuum insulation panels in the basement.</li></ul>	
Floors	Not insulated.	
Roof	280mm cellulose insulation.	
Windows	Triple glazed to *PassivHaus* standards.	
External doors	No change.	
Other sustainability features	Unknown.	
Airtightness	Unknown.	
Ventilation	Whole-house ventilation system with heat recovery.	Heat recovery: 92%
Heating	Wood pellet boiler with solar heating.	
Renewable energy sources	8m² evacuated tube solar-thermal system for space and water heating.	

Change in Energy Use and CO_2 Emissions

Table 4.25 Change in energy use before and after refurbishment in Case Study 4.12

	Before refurbishment	After refurbishment	Savings
Total primary energy used	115,500kWh/year	5250kWh/year	95% (110,250kWh/year)
CO_2 emissions			95% (29 tonnes CO_2/year)

Costs

Not available.

Planning Issues

The external appearance of the house was essentially unchanged despite the application of external insulation.

Conclusions

The owners believe that they have a top-quality living environment with low consumption costs. The house has created great interest in the neighbourhood and has made energy-efficient refurbishment a topic of conversation with the owners. The house was included in the Deutsche Energie-Agentur *Efficient Homes* project in 2003.

Figure 4.56 Interior after refurbishment

Source: DENA

Case Study 4.13 Germany: Eichstetten

- 300-year-old half-timbered house (a listed building).
- Refurbishment brings together the traditional and modern.
- 83 per cent primary energy savings.
- Several different wall insulation systems were used according to their possibilities.
- Heating supplied by large solar system and wood pellet boiler.

Figure 4.58 The interior before refurbishment

Source: DENA

Figure 4.57 The house before refurbishment

Source: DENA

Description

This 300-year-old building is in Eichstetten, south Germany, near the French border. Like many in the area, the half-timbered building is listed and the refurbishment thus had to respect conservation practices for listed buildings. The 260m² house has three storeys and a basement.

Figure 4.59 The house after refurbishment

Source: DENA

Refurbishment Process

The house was substantially renovated by the owners in order to bring it up to date, with the aim of uniting the traditional and the modern through an ecological approach. The objective was a comprehensive energy renewal that complied with the latest technical standards, but within conservation practices. Careful design and implementation of the insulation measures were necessary, with different methods used in different parts of the building. The existing windows were retained in order to keep the same appearance, but draught-stripped. The house is heated by a large solar thermal array and a wood pellet boiler. Natural materials such as loam render and solid wood flooring were combined with a modern ventilation system, providing a pleasant indoor climate.

Figure 4.60
Ventilation system in loft space

Source: DENA

Sustainability Improvements

Table 4.26 Sustainability improvements in Case Study 4.13

Walls	• Front gable: 160mm of cellulose insulation injected into the cavity between the façade and newly mounted internal boards. • Other façades: 60mm and 80mm wood fibre and cellulose batts. • Basement: mineral fibre or cellulose on walls.
Floors	Not insulated.
Roof	200mm cellulose insulation.
Windows	Single glazed with new draught-stripping in wood frames.
External doors	Unknown
Other sustainability features	Use of natural materials (e.g. loam rendering and solid wood flooring).
Airtightness	Not tested.
Ventilation	Central ventilation system with heat recovery. Heat recovery: >90%
Heating	Wood pellet boiler and solar panels.
Renewable energy sources	13m² solar thermal panels for space and water heating.

Comfort and Overheating

The owners are very proud of the results, and a very pleasant year-round indoor climate has been produced.

Change in Energy Use and CO_2 Emissions

Table 4.27 Change in energy use before and after refurbishment in Case Study 4.13

	Before refurbishment (calculated)	After refurbishment (calculated)	Savings
Total primary energy used	52,338kWh/year	9048kWh/year	83% (43,290kWh/year)
CO_2 emissions			83% (20 tonnes CO_2/year)

Costs

Not available.

Planning Issues

Listed building status meant that the external appearance had to remain unchanged and the walls had to be insulated internally.

Conclusions

The owners are now happy that they have made their 300-year-old house very up to date. The building combines all of the internal and external historic features of the building with modern services. The house was included in the Deutsche Energie-Agentur *Efficient Homes* project in 2003.

Figure 4.61 Interior after refurbishment

Source: DENA

Case Study 4.14 Germany: Constance

- 1950s lakeside house fully renovated.
- Larch cladding provides new contemporary appearance.
- 86 per cent energy savings.
- 300mm of insulation in all elements.
- Triple-glazed windows.
- Heat pump used for heating with solar collectors.
- House refurbished for rental.

Description

This timber-clad house is in an idyllic setting on Lake Constance, a 540km² lake bordering Germany, Austria and Switzerland, near the Alps. The house was built by the current owners in 1959 and extended to include a granny flat during the 1960s. The whole house was in poor condition and the owners decided that it needed a complete renovation and installation of energy-efficient technology in order to maintain its value and enable it to be rented out on a long-term basis.

Refurbishment Process

There was an urgent need to install a new heating system, to improve the internal environment and to replace the kitchen and bathroom. At the same time, the owners decided to give the house a contemporary appearance with a new homogeneous wood façade and large windows facing onto the lake. The existing walls were clad in a new timber frame and 300mm of insulation was installed behind a larch cladding. An equal thickness of insulation was installed in the roof by adding a new frame to the roof timbers, and in the cellar. New triple-glazed windows were used and heating provided by an air-to-water heat pump and a solar collector system.

Figure 4.62 The house before refurbishment

Source: DENA

Figure 4.63 The house after refurbishment

Source: DENA

Figure 4.64 Flat-plate solar collector panels

Source: DENA

Sustainability Improvements

Table 4.28 Sustainability improvements in Case Study 4.14

Walls	<ul><li>Front façade: 300mm injected cellulose insulation behind a new timber frame.</li><li>Extension: 300mm of extruded polystyrene applied externally, with cladding.</li></ul>
Floors	300mm polyurethane foam insulation.
Roof	300mm blown cellulose fibre insulation in new frame.
Windows	Triple-pane insulating glass.
External doors	As windows.
Other sustainability features	Unknown.
Airtightness	Unknown.
Ventilation	Central ventilation system with heat recovery. Heat recovery > 80%
Heating	Central air-to-water heat pump supported by a solar collector system.
Renewable energy sources	4.1m² flat-plate solar-collector system for space and water heating.

Figure 4.65 Ventilation intake and exhaust

Source: DENA

Comfort and Overheating

The indoor climate is now considered excellent, so that even time spent in the small bedrooms is pleasant, even with the windows closed. The internal ambiance is now praised by family friends and visitors.

Change in Energy Use and CO$_2$ Emissions

Table 4.29 Change in energy use before and after refurbishment in Case Study 4.14

	Before refurbishment (calculated)	After refurbishment (calculated)	Savings
Total primary energy used	52,710kWh/year	7224kWh/year	86% (45,486kWh/year)
CO$_2$ emissions			86% (30 tonnes CO$_2$/year)

Costs

Not available.

Planning Issues

None reported.

Conclusions

The owners are very pleased with the whole project and their feeling of well-being is so much greater due to the new indoor climate. A new tenant for the house was found soon after refurbishment. The house was included in the Deutsche Energie-Agentur *Efficient Homes* project in 2003.

Figure 4.66 View from inside the house

Source: DENA

Case Study 4.15 UK: London (Kings Cross)

- Small top-storey apartment in central London, sustainably refurbished and providing 64 per cent CO_2 savings.
- Space-heating energy consumption reduced by 82 per cent.
- Innovative ceiling insulation used.
- Mixture of secondary and new double glazing.
- Wall-mounted condensing combination boiler increases usable space.
- Heating radiators moved from external walls.

Figure 4.67 The apartment (top right) has a view of St Pancras Station

Source: Simon Burton

Description

This 67m² top-floor corner flat (on the seventh floor) in central London was very cold in winter due to solid brick walls, an uninsulated roof and single-glazed windows. The flat, built during the early 1900s, had been rented out since it was purchased by the owners in 2002, and while minor improvements had been carried out over the years, it was decided to carry out major work before the owners made it their own residence. The original ceilings showed the roof construction of steel joists with infill clinker concrete with no mouldings and many cracks. The solid walls had also moved and cracked over the years. The windows were mainly single-glazed sash windows, with a circular-topped metal-framed window in the sitting room. Heating was supplied by an old floor-mounted boiler in the kitchen.

Refurbishment Process

The refurbishment had multiple aims: to reduce energy consumption, to improve comfort, to optimize usable space, and to improve appearance. The original cast-iron floor-mounted boiler was replaced by a wall-mounted, condensing combination boiler, giving more floor area in the kitchen and more cupboard space by removing the hot water cylinder and pressure tanks. The solid external walls were lined with 82.5mm thermal board, fixed with plaster dabs and mechanical plugs. This thickness was chosen as optimal in terms of loss of internal space against insulation value. The flat roof was believed to be basically uninsulated, and being only a part of the large roof area of the block, external insulation of a single flat was not seen as feasible. The solution adopted was to fix internal insulation over the whole ceiling, making sure that it provided an effective vapour barrier to stop internal condensation on the (now) cold roof. Timber battens were fixed to the ceiling and the space between filled with polyisocyanurate board, with a final layer of thermal board with phenolic foam insulation. Joints between the thermal board were filled with expanding polyurethane foam to complete the vapour barrier. The windows on the road frontage could not be replaced due to the need for scaffolding, so secondary glazing has been fitted. The large sash window in the kitchen, which makes up the larger part of the exposed kitchen wall, has been replaced by a new double-glazed sash window.

In order to upgrade the internal appearance, new plaster ceiling mouldings were fitted in the living and bedrooms, with careful treatment where the ceiling had been lowered over windows. The central heating radiators previously placed under the windows, fed by ugly long pipework fixed to the skirting boards, have been replaced by smaller radiators clustered in the central part of the flat on the basis that the insulated walls and secondary-glazed windows mitigate the need to have radiators on external walls.

Figure 4.68 Expanding polyurethane foam was used to seal between the ceiling insulation boards

Source: Ray Unwin

Sustainability Improvements

Table 4.30 Sustainability improvements in Case Study 4.15

Walls	Internal insulation: 70mm polyisocyanurate, bonded to plasterboard.	U value: 0.24W/m²K
Floors	Internal: no change.	
Roof	Internal ceiling insulation: 30mm polyisocyanurate plus 25mm phenolic foam, bonded to plasterboard.	U value: 0.25 W/m²K
Windows	• Road frontage: secondary glazing. • Access walkway: double-glazed, argon-filled, low e-coating sash window.	
External doors	Unchanged.	
Other sustainability features	None.	
Airtightness	All windows and front door draught-stripped.	Not tested
Ventilation	Extract fan in bathroom (to be fitted in kitchen).	
Heating	Condensing combination gas boiler.	
Renewable energy sources	None.	

Comfort and Overheating

The occupants are delighted with the warmth of the flat: it heats up quickly, retains the heat and the boiler rarely runs after heating up. The internal appearance is greatly improved, particularly the ceilings and the cracked external walls, complemented by the new mouldings. There remains plenty of thermal mass in the internal brick walls and it is thought that the insulation, particularly that on the ceiling, will stop the overheating of the flat during summer. The small loss of floor space from the wall insulation is offset by the greater storage space from the loss of floor-mounted boiler and storage tanks.

Figure 4.69 The new double-glazed, spring-balanced sash window

Source: Simon Burton

Change in Energy Use and CO_2 Emissions

Table 4.31 Change in energy use before and after refurbishment in Case Study 4.15

	Before refurbishment (calculated)	After refurbishment (calculated)	Savings
Space heating	20,300kWh/year	3700kWh/year	82% (16,600kWh/year)
Water heating	5170kWh/year	2250kWh/year	56% (2920kWh/year)
Electricity used	2250kWh/year	1970kWh/year	12% (280kWh/year)
Total CO_2 emissions	6.1 tonnes CO_2/year	2.2 tonnes CO_2/year	64% (3.9 tonnes CO_2/year)
Cost savings (approximate)			UK£750/year

Costs

The whole refurbishment of the apartment has cost just under UK£20,000 with the insulation of the external walls and ceilings costing UK£7700 (plus VAT), and the new double-glazed window UK£2200. This is a fraction of the value of a central London flat of this type.

Planning Issues

There were no planning issues. The apartment block is not in a conservation area and is not a listed building. No change to the external appearance was necessary.

Conclusions

Cold bridging where the internal walls meet the external walls and the roof and protruding chimneys is inevitable, and the small size of the flat would prohibit more insulation of these elements. Careful inspection will be maintained to check for condensation and mould growth. Generally, keeping humidity levels low is important and an extract system still needs to be fitted in the kitchen to complement that in the bathroom. The occupants are very pleased with the whole refurbishment, which demonstrates what can be done at low cost in small city apartments to improve comfort and greatly reduce heating energy use.

Figure 4.70 New mouldings improve the internal appearance

Source: Simon Burton

5 Good and Best Practice in Apartment Blocks and Social Housing

A great deal of housing accommodation, particularly in towns, occurs in groups of apartments or houses, which may be low or high rise. Both social and private landlords, as well as owners and leaseholders, have been responsible for undertaking sustainable refurbishment to complete blocks or terraces, though as with single houses, the drivers and processes are likely to vary.

There are several drivers for making social housing more sustainable, and this has produced many examples of good and best practice in different countries. Social housing is normally seen as a service to provide poorer households with affordable housing, and fuel poverty and fuel bills can be very important issues. Government and charity funding may be available for innovative approaches and works, providing long-term sustainability benefits.

There are several examples where private owners have sustainably upgraded whole housing blocks to provide high-quality apartments for renters and long-term leaseholders, with the objective of attracting or keeping tenants, or increasing rents and property values.

A third situation is where a block of housing is privately owned by a number of occupiers and comprehensive renovation has been possible by the owners, forming an association or housing co-operative and agreeing to common refurbishment.

Refurbishing a block or group of housing has the advantage of incorporating economies of scale within design and construction; but it may be difficult to coordinate and get agreement from all occupants on the best universally applicable solutions. Extensive consultation and discussion of options with occupiers have achieved good results in many situations.

The case studies in this chapter include examples of good practice where general refurbishment focusing on energy efficiency has been a major consideration, as well as projects where *PassivHaus* standards have been the target. Again, the choice does not represent the result of any cross-country survey: these examples have been chosen to demonstrate sustainable refurbishment in a range of countries, building types and locations, where a range of sustainability options have been used.

Table 5.1 The case studies

Case studies				Restrictions	Sustainability solutions						
Case study number	Type	Location	Age		Wall insulation to existing walls	Roof insulation	Floor insulation	Ventilation	Windows	Heating system	Renewables
5.1	Large East German estate	Germany: Leipzig	1973	None	External: 100mm mineral wool insulation slabs	120mm insulation in the top-floor ceiling	80mm insulation to the basement ceiling	Extract fans in each flat	Double-pane insulated glass	District heating based on combined heat and power	181m² of solar thermal panels
5.2	Four two-storey blocks of apartments	Germany: Ahrensburg	1976	None	External: 180mm insulation	220mm cellulose insulation	110mm of insulation to basement ceilings	Central extract system	Triple-glazed insulating glass	District heating from wood chip boiler	120m² of solar collectors
5.3	Historic house with six apartments in the city	Germany: Mannheim	1901	Listed building	Internal: 100mm insulation	Rebuilt with insulation to achieve low-energy housing standard (KfW40)	100mm expanded polystyrene in the basement	Whole house with 90% heat recovery	Double-glazed glass	Local district heating network	None
5.4	Block of six apartments in a village	Germany: Hopferau	1963	None	Existing: 360mm; extension: 240mm Attic: 120mm	360mm wood fibre	80mm insulation to basement ceiling	Whole-house ventilation system with 80% heat recovery	Triple-pane insulated glazing	Central wood pellet boiler	20m² of solar collector panels
5.5	Semi-detached houses in town	UK: St Albans	1930s	Permitted development	External insulation with 70mm phenolic foam	270mm of insulation to rear extension; loft spaces topped up to 270mm	None	Natural ventilation; trickle vents over windows	Double glazed with 'edgetech' super-saver frames	Gas boilers	2.75m² of flat-plate solar thermal collectors per house

5.6	High-rise block	UK: Smethwick	1960s	None	External insulation: 80mm mineral fibre	Insulation under profiled aluminium roof	None	Positive pressure input and extract fans	Double glazed with argon fill in composite aluminium timber frames	Gas combination boilers replaced electric heating	None
5.7	Four-storey block	Austria: Sankt Florian	1982	None	External insulation	Insulation added in the pitched roof	Insulation of the cellar ceiling	Natural ventilation	Triple glazed	Gas central heating	None
5.8	Low-rise block	Bulgaria: Sofia, Zaharna Fabrika	1947	None	External insulation of solid brick walls	Insulation of the new roof	Insulation of the basement ceiling		Double-glazed windows in PVC frames	District heating system	None
5.9	High-rise block	Slovenia: Ljubliana	1965	None	External insulation	Insulation added	No change		Double glazed, argon fill, low e, insulated frames		
5.10	Low-rise block	Sweden: Alingsås	1970	None	Long sides: 480mm external insulation; ends: 130mm insulation	300mm of loose mineral wool insulation	40–60mm of cellular concrete with 60mm expanded polystyrene	Whole-house ventilation with 80% heat recovery	Triple glazed, low e with krypton, frame highly insulated	Heating coil in the ventilation systems from district heating	None at present
5.11	Terraced housing	The Netherlands: Raamsdonk	1963–1969	None	External insulation, including insulated cladding bricks	Insulation added to uninsulated roofs	No action	Mechanical ventilation with 80% heat recovery	High-efficiency double glazing	Gas-fired boilers	Some units with solar collectors

Case Study 5.1 Germany: Leipzig

- Large block of flats in former East Germany renovated.
- Converted from an unfavourable area to one in which demand exceeds supply.
- External insulation transforms precast concrete block.
- 75 per cent primary energy savings achieved.
- Heating from a district heating system based on combined heat and power plant.
- Photovoltaic (PV) cladding to balconies.

Description

This block of 167 flats built in 1973 is in Leipzig, in former East Germany. It was a traditional prefabricated concrete block and had become an unfavourable area in which to live. The property-owning company decided to carry out an energy-efficient refurbishment together with other improvements, and this has transformed the block into one in which demand exceeds supply.

Refurbishment Process

Comprehensive improvements were carried out to the whole block, including external insulation on the walls and roof, and basement insulation, combined with double-glazed windows. New balconies were added and the green spaces around the block were renovated. Heating is supplied by a district heating system based on a combined heat and power plant, and solar thermal panels are integrated within the balconies.

Figure 5.1 Block before refurbishment

Source: DENA

Figure 5.3 Balconies incorporating photovoltaic arrays

Source: DENA

Figure 5.2 Block after refurbishment

Source: DENA

Sustainability Improvements

Table 5.2 Sustainability improvements in Case Study 5.1

Walls	External: 100mm mineral wool insulation slabs.
Floors	80mm insulation to the basement ceiling.
Roof	120mm insulation in the top-floor ceiling.
Windows	Double-pane insulated glass.
External doors	No change.
Other sustainability features	Unknown.
Airtightness	Not tested.
Ventilation	Individual extract system in each flat.
Heating	District heating based on combined heat and power plant.
Renewable energy sources	181m^2 of solar thermal panels.

Comfort and Overheating

The whole living atmosphere in the block has been completely enhanced and the tenants benefit from greatly reduced energy bills for heating and hot water.

Change in Energy Use and CO_2 Emissions

Table 5.3 Change in energy use before and after refurbishment in Case Study 5.1

	Before refurbishment (calculated)	After refurbishment (calculated)	Savings
Primary energy used per flat	11,371kWh/year	2843kWh/year	75% (8528kWh/year)
CO_2 emissions per flat			75% (2.6 tonnes CO_2/year)

Costs

Not available.

Planning Issues

The appearance of the block and area has been greatly improved.

Conclusions

The property now holds its own in the market and the company has improved its image as a result of the renovation. Service charges have been lowered and the quality of life on the estate has increased. Both the owners and the tenants have benefited from the refurbishment.

Figure 5.4 The estate has been transformed

Case Study 5.2 Germany: Ahrensburg

- Four blocks of 1970s housing transformed in appearance and energy efficiency.
- Balconies and a third complete storey added.
- Thick external insulation, with roof and basement insulation and triple-glazed windows.
- Heating from wood chip district heating system.
- 90 per cent primary energy savings.
- Benefits to owners and occupants recognized.

Description

Ahrenburg is a small town near Hamburg in northern Germany. This development of four blocks of two-storey flats was built in 1976; the owner decided that a complete renovation was necessary, together with an extension to add a third storey. The development now has 27 flats in the four renovated blocks.

Refurbishment Process

The appearance of the blocks has been transformed by the addition of balconies to the original buildings and the construction of a third-storey set back with spacious roof terraces. The owner chose the option with the best cost–benefit ratio and also decided to apply a low-energy house standard. This included thick external insulation, loft insulation and triple-glazed windows. Heating is provided by a local district heating system run on wood chips with solar collectors contributing to the hot water supply.

Figure 5.5 One block before refurbishment

Source: DENA

Figure 5.6 Blocks after refurbishment

Source: DENA

Figure 5.7 180mm of wall insulation and triple-glazed windows

Source: DENA

Sustainability Improvements

Table 5.4 Sustainability improvements in Case Study 5.2

Walls	External: 180mm insulation.
Floors	110mm of insulation to basement ceilings.
Roof	220mm cellulose insulation.
Windows	Triple-glazed insulating glass.
External doors	No change.
Other sustainability features	Unknown.
Airtightness	Unknown.
Ventilation	All flats connected to a central extraction system.
Heating	District heating system run on wood chips with solar hot water.
Renewable energy sources	A total of 120m^2 of solar thermal collectors for domestic hot water.

Comfort and Overheating

Greatly improved indoor air quality is seen as one of the main benefits to occupants, together with an improved layout.

Change in Energy Use and CO$_2$ Emissions

Table 5.5 Change in energy use before and after refurbishment in Case Study 5.2

	Before refurbishment (calculated)	After refurbishment (calculated)	Savings
Primary energy used per flat	14,670kWh/year	1385kWh/year	90% (13,284kWh/year)
CO$_2$ emissions per flat			90% (15 tonnes CO$_2$/year)

Costs

Not available.

Planning Issues

No difficulties with planning permission are recorded.

Conclusions

The occupants of the refurbished blocks are reputed not to remember what the original blocks once looked like. Demand for the now highly efficient units has soared despite higher prices.

Figure 5.8 The residents appreciate the quality of the refurbishment

Source: DENA

Case Study 5.3 Germany: Mannheim

- Historic building brought up to date in comfort and energy-use terms.
- External appearance unchanged.
- 86 per cent energy savings produced.
- Internal wall insulation combined with basement and roof insulation.
- Modern energy standards incorporated within a listed building.

building built in 1901 on a street dominated by historic buildings. There are six separate flats in the house. Refurbishment had to take into account the historic nature of the building, together with high standards of energy efficiency and comfort.

Refurbishment Process

The owners attached great importance to combining an energy-efficient refurbishment with the conservation criteria for listed buildings. This has meant internal wall insulation, basement and roof insulation, and double-glazed windows, combined with heat-recovery ventilation systems in each flat. Heating comes from a district heating network.

Figure 5.9 The building before refurbishment

Source: DENA

Figure 5.10 The building after refurbishment

Source: DENA

Description

Mannheim is a city of more than 300,000 inhabitants in south-west Germany, not far from the French border. This case study involves a historic

Sustainability Improvements

Table 5.6 Sustainability improvements in Case Study 5.3

Walls	Internal insulation: 100mm.	
Floors	100mm expanded polystyrene in the basement.	
Roof	Rebuilt with insulation to achieve low-energy housing standard (KfW40) in Germany.	
Windows	Double glazed.	
External doors	No change.	
Other sustainability features	Unknown.	
Airtightness	Unknown.	
Ventilation	Whole-house ventilation system with heat recovery.	Heat recovery: 90%
Heating	Connected to the local district heating network.	
Renewable energy sources	None.	

Comfort and Overheating

The occupants now enjoy not only the charm of this century-old dwelling, but also the comfort of modern energy standards with a comfortable indoor climate throughout and low energy consumption and costs.

Figure 5.11 The ventilation system controller

Source: DENA

Change in Energy Use and CO_2 Emissions

Table 5.7 Change in energy use before and after refurbishment in Case Study 5.3

	Before refurbishment (calculated)	After refurbishment (calculated)	Savings
Primary energy used per flat (average)	30,225kWh/year	4417kWh/year	86% (25,807kWh/year)
CO_2 emissions per flat			86% (5.3 tonnes CO_2/year)

Costs

Not available.

Planning Issues

This is a listed building in a historic street and the refurbishment was carried out without changing the external appearance of the street façade.

Conclusions

Since the refurbishment, the property has been considered a real jewel, not only to look at but also from an energy and urban planning perspective. The result is that the street's character has been maintained for the future.

Figure 5.12 The historic character of the house has been maintained

Source: DENA

Case Study 5.4 Germany: Hopferau

- 1960s apartment block in southern Germany given a completely new appearance.
- Prefabricated roof and façade elements included high insulation levels.
- Near *PassivHaus* standards achieved.
- Triple-glazed windows.
- Heating from a central wood pellet boiler.

Description

Hopferau is a small village and municipality in the Alpine foothills of south Germany, near the border with Austria. The six-apartment block was built in 1963 in a 'regional style' and the owners decided to upgrade it both to give it a modern appearance and greatly improve sustainability.

Refurbishment Process

The approach taken by the owners was to use innovative prefabrication techniques, a practice that had long been employed successfully for houses constructed of timber. Façade and roof elements made of timber were fabricated in the factory and included high levels of insulation. This was believed to be a much cheaper process than conventionally installing insulation on the building site. At the same time, a conservatory extension was added on the south side and the roof of the original building was raised to provide more living space.

Figure 5.13 The block before refurbishment

Source: DENA

Figure 5.14 The block after refurbishment

Source: DENA

Figure 5.15 The windows are triple glazed

Source: DENA

Sustainability Improvements

Table 5.8 Sustainability improvements in Case Study 5.4

Walls	Existing walls: 360mm of wood fibre and mineral wool.
	New extension: 240mm of mineral wool.
	Attic storey: 120mm wood fibreboard.
Floors	80mm insulation to basement ceiling.
Roof	360mm wood fibre above and between rafters.
Windows	Triple-pane insulated glazing.
External doors	No change.
Other sustainability features	Unknown.
Airtightness	Unknown.
Ventilation	Whole-house ventilation system with heat recovery. Heat recovery < 80%
Heating	Central wood pellet boiler plus solar collectors.
Renewable energy sources	20m^2 of solar collector panels.

Comfort and Overheating

Benefits were immediately apparent to the occupants: the insulation, the conservatory extensions and the improved integration of daylighting all made for a very comfortable indoor climate. The heating rarely needs to be turned on.

Change in Energy Use and CO$_2$ Emissions

Table 5.9 Change in energy use before and after refurbishment in Case Study 5.4

	Before refurbishment (calculated)	After refurbishment (calculated)	Savings
Total primary energy used per flat	26,258kWh/year	2593kWh/year	90% (23,664kWh/year)
CO$_2$ emissions per flat			90% (8.3 tonnes CO$_2$/year)

Costs

Not available.

Planning Issues

There were no planning issues.

Conclusions

The owners claim that they carried out a truly show-case project, achieving a considerable reduction in energy costs and greatly improving the comfort and well-being of the occupants.

Figure 5.16 Front door treatment

Source: DENA

Case Study 5.5 UK: St Albans

- Difficult to heat housing estate was brought above building regulation standards.
- External insulation with phenolic foam and render.
- Roofs were insulated.
- Solar panels for hot water.
- Double-glazed windows installed.

Figure 5.18 Typical house after refurbishment

Source: St Albans CDC

Figure 5.17 Houses before refurbishment

Source: St Albans CDC

Description

St Albans is in south-east England just north of London. The Marford Road project was undertaken by St Albans City and District Council, focusing on the transformation of 18 'hard-to-treat' homes into an 'eco-street'. The semi-detached houses form part of the council's social housing stock. The project aimed to bring the 70-year-old properties in line with the Energy Saving Trust's best practice solution, giving a 25 per cent improvement in energy efficiency according to the then current building regulations. Before the improvements, residents found the houses cold, with frost on the insides of the windows and regular condensation and mould problems.

Refurbishment Process

The houses had been identified as being below the UK government's Decent Homes Standard and substandard in terms of heating. Work on the refurbishment started in May 2007.

Figure 5.19 Installing the solar panels

Source: St Albans CDC

Figure 5.20 Insulated back door

Source: St Albans CDC

Sustainability Improvements

Table 5.10 Sustainability improvements in Case Study 5.5

Walls	External insulation with Alumasc system, 70mm phenolic foam with polymer modified through colour render.	U value: 0.35W/m²K
Floors	No action taken.	
Roof	Loft spaces already insulated; if not, topped up to 270mm. 270mm of insulation added to the uninsulated rear extension roofs.	U value: 0.16W/m²K
Windows	Double-glazed with 'edgetech' super-saver frames.	U value: 1.5W/m²K
External doors	Solid doors with insulation.	U value: 1W/m²K
	Half-glazed doors.	U value: 1.5W/m²K
Other sustainability features	None.	
Airtightness	Improved.	5m³/m²/hour
Ventilation	Natural, with over glass vents in windows.	
Heating	Existing boilers retained if under ten years old and condensing.	
Renewable energy sources	2.75m² of flat-plate solar thermal collectors fitted flush to new roofs.	Up to 55% of annual hot water

Comfort and Overheating

The residents find that the houses are much warmer with the external insulation, and the solar panels often provide all of the water heating needs during summer.

Change in Energy Use and CO_2 Emissions

Table 5.11 Change in energy use before and after refurbishment in Case Study 5.5

	Before refurbishment	After refurbishment	Savings
CO_2 emissions	Not available	Not available	2 tonnes CO_2/year
Cost savings (approximate)			UK£300/year per house

Costs

The cost of the works was approximately UK£22,000 per property. The project was considered to be high cost, but intended as an exemplar project and the economies of scale made the works affordable.

Planning Issues

There were no planning issues as all of the improvements were 'permitted development'.

Conclusions

After completion of the works, the residents were supplied with a brochure explaining the improvements and giving 'Top Tips' and information on energy-saving behavioural changes, such as turning their washing machine dial to 30°C and only boiling the amount of water required.

Figure 5.21 Infrared photography shows the cold bridge of the lintel over the back door

Source: St Albans CDC

Case Study 5.6 UK: Smethwick

- 1960s high-rise block refurbished.
- Clad with 80mm external insulation.
- New timber-framed double-glazed windows with powdered aluminium covering.
- Electric heating replaced with gas combination boilers.
- Extensive tenant consultation.

Description

Thompson Gardens is a 15-storey block of flats built during the early 1960s in Smethwick near Birmingham, UK. It is owned by Sandwell Metropolitan Borough Council and has a total of 89 flats, a mixture of one and two bedrooms, all of which have an external face on the front/rear elevations. It is constructed with a reinforced concrete frame with infill cavity brick/block panels and was structurally considered to be in fair to good condition, but in need of external maintenance. Flats had electric space and water heating, though gas was available for cooking. The block of flats was seen as very popular; but the safety and environment needed improvement.

Refurbishment Process

The structure was surveyed in 2005 in order to decide what needed to be done to bring it up to the UK Government's Decent Homes Standard. Exten-

Figure 5.22 Block before refurbishment

Source: Sandwell MBC

Figure 5.23 Block after refurbishment

Source: Sandwell MBC

sive consultation was carried out with the tenants concerning the treatment of balconies and the colour choice for the block. The residents remained in their flats during the refurbishment and a temporary respite centre was provided on the site to provide a quiet area for residents wishing to escape the noise of the building works. The refurbishment scheme was designed to be both functional and aesthetically pleasing, and to use recyclable materials or materials from sustainable sources. A further objective was to use low-maintenance materials.

As well as relatively minor structural repairs and the insulation works, the balconies were enclosed and improvements were made to the ground-floor entrance; a new entry phone system was also installed and the surrounding area was landscaped. The over-cladding solution was seen to solve many ongoing problems of water penetration, structural deterioration and high heat loss, and would also help by transferring wind loads to the floor plates, thus eliminating the need for further (or replacement) wall ties. The refurbishment would make the building maintenance free for a 30-year period.

Sustainability Improvements

Table 5.12 Sustainability improvements in Case Study 5.6

Walls	External insulation: 80mm mineral fibre with basecoat, primer and a 1.5mm silicone texture coating; brickwork cladding at ground level.	U value: $0.35W/m^2K$
Floors	No change.	
Roof	New profiled aluminium roof with insulation.	
Windows	Double glazed with argon fill, in softwood timber frames with polyester powder-coated aluminium exterior covering.	
External doors	No change.	
Other sustainability features	Use of recyclable materials and materials from sustainable sources.	
Airtightness	Draught-stripping.	
Ventilation	Positive-pressure input system extract fans in kitchens and bathrooms; use of low watt variable volume fans.	
Heating	New combination gas boilers for space heating and hot water.	
Renewable energy sources	None.	

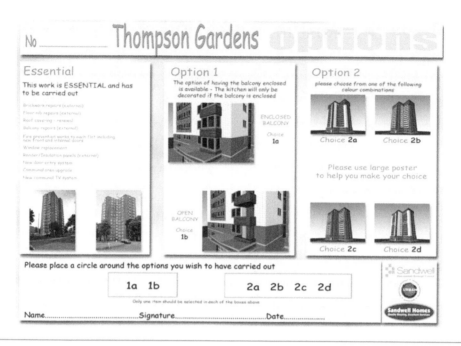

Figure 5.24 Tenant consultation

Source: Sandwell MBC

Change in Energy Use and CO$_2$ Emissions

Table 5.13 Change in energy use before and after refurbishment in Case Study 5.6

Per flat on average	Before refurbishment (calculated)	After refurbishment (calculated)	Savings
Heating, DHW and lighting used	163kWh/year/m^2	97kWh/year/m^2	40% (66kWh/year/m^2)
CO$_2$ emissions	4.1 tonnes CO$_2$/year	1.4 tonnes CO$_2$/year	66% (2.7 tonnes CO$_2$/year)
Cost savings (approximate)			UK£815/year

Costs

Not available.

Planning Issues

No issues were involved.

Conclusions

Many of the residents have commented on how comfortable the block now is to live in and are very enthusiastic at how warm the flats have become; they have been able to turn their heating down (and, in some cases, off) much earlier than they were able to in past years.

Case Study 5.7 Austria: Sankt Florian

- Housing co-operative carries out comprehensive refurbishment.
- Agreement reached on refurbishment measures after options survey of owners.
- Insulation of walls, floors and roof.
- Triple-glazed windows.
- 66 per cent energy saving achieved.

Description

This apartment block in upper Austria contains 48 apartments, the owners of which form a social housing association. The block was built in 1982 and the owners requested renovation as the outside doors and window areas were old and partly damaged, making living in the building uncomfortable.

Refurbishment Process

The overall aim was to implement a comprehensive renovation, not only to replace damaged parts of the building. As the building is jointly owned by the apartment owners, all of them had to agree to the renovation measures, and originally they differed about the renovation measures, some only wanting a 'small renovation', while others wanted a comprehensive renovation of the building. The housing association organized a written survey where the apartment owners could choose between three options. As a result of the survey, a general and comprehensive renovation was agreed upon and implemented in 2005.

Figure 5.26 The block after renovation

Source: EI-Education

Figure 5.25 The block before renovation

Source: EI-Education

Sustainability Improvements

Table 5.14 Sustainability improvements in Case Study 5.7

Walls	External insulation.	U value: 0.17–0.24W/m²K
Floors	Insulation of the cellar ceiling.	U value: 0.22W/m²K
Roof	Insulation added in the pitched roof.	U value: 0.11W/m²K
Windows	Triple-glazed windows.	U value: 1.3W/m²K
External doors	Insulated.	U value: 1.3W/m²K
Other sustainability features	Renovation of entrance areas and roof.	
Airtightness	Unknown.	
Ventilation	Unknown.	
Heating	Gas central heating.	
Renewable energy sources	None.	

Comfort and Overheating

The energy-saving potential is very high but also depends on user behaviour. With the new windows and insulated outside walls, the owners have to learn new habits (e.g. in relation to heating, to airing and when to close the windows and sun blinds during summer). This is very important in order to fully achieve the energy-saving potential.

Change in Energy Use and CO_2 Emissions

Table 5.15 Change in energy use before and after refurbishment in Case Study 5.7

	Before refurbishment (calculated)	After refurbishment (calculated)	Savings
Gas used (average per flat)	10,604kWh/year	3569kWh/year	66% (7035kWh/year)

Costs

The total cost of the renovation was 1,215,000 Euros, shared amongst the owners.

Planning Issues

No planning issues were reported.

Conclusions

The renovation phase was difficult for the people living in the building. There were problems with dust and dirt; but in the end, when the renovation was finished, all owners and tenants were happy and satisfied with their 'new' homes.

Case Study 5.8 Bulgaria: Sofia, Zaharna Fabrika

- Block of apartments owned by the occupants.
- Insulation of walls, floors and ceilings.
- Double-glazed windows installed.
- Two new flats added in the roof space.
- Rent from the new flats paid off half the loan.
- Innovative financing to take account of low- and middle-income occupants.
- 61 per cent space heating reduction.

Description

This multi-dwelling building containing 13 flats was built in 1947. All flats are owner occupied and the owners formed the first association of owners to be registered in Bulgaria. The refurbishment was carried out in 2004.

Refurbishment Process

The objective of the project was to renovate and carry out further maintenance of the multi-dwelling building in which the flats are owned by the inhabitants. The major issues solved include overcoming the problems that arise from the low incomes of the owners and their different interests. The renovation was also intended to lead to lower energy consumption, as well as improvements to the comfort of the flats.

The project also included the complete reconstruction of the roof and conversion of the attic, which consisted of two common premises, into small flats. The rent of these new flats was intended to help in repaying the loan.

Figure 5.27 The building before renovation

Source: EI-Education

Figure 5.28 The building after renovation

Source: EI-Education

The project was undertaken by the Bulgarian Housing Association in partnership with the De Nieuwe Unie Housing Association, Rotterdam, and the Woondrecht Housing Association, Dordrecht, The Netherlands.

Sustainability Improvements

Table 5.16 Sustainability improvements in Case Study 5.8

Walls	External insulation of solid brick walls.
Floors	Insulation of the basement ceiling.
Roof	Insulation of the new roof.
Windows	Double-glazed windows in PVC frames.
External doors	Unknown.
Other sustainability features	Unknown.
Airtightness	Unknown.
Ventilation	Natural.
Heating	District heating system, pipework insulated and system balanced.
Renewable energy sources	None.

Figure 5.29 The loft space being converted into flats

Source: EI-Education

Comfort and Overheating

The inhabitants are reported to be satisfied by the results as the insulation of the external envelope has resulted in better internal comfort, together with large energy savings.

Change in Energy Use and CO_2 Emissions

Table 5.17 Change in energy use before and after refurbishment in Case Study 5.8

	Before refurbishment (calculated)	After refurbishment (calculated)	Savings
Space heating per flat	13,758kWh/year	5372kWh/year	61% (8386kWh/year)

Costs

The total project cost was 52,375 Euros, which was financed through a loan from banks in The Netherlands, who offered lower interest rates; the loan is for 20 years. The monthly payment of the loan is approximately 350 Euros; but half of this amount comes from the rent of the two new flats in the attic.

Planning Issues

No issues were raised.

Conclusions

The residents are happy that the renovation has lengthened the life span of the building by 40 years. The conversion of the roof space into flats has been a very important element in the project's funding, and it is thought that most of the buildings in the area could be similarly extended with an additional floor. The importance of flexibility from the financing institutions was very important as most of the owners were on low or medium incomes.

Case Study 5.9 Slovenia: Ljubljana

- Large block renovated with options for owners.
- External insulated cladding.
- Insulation to the roof.
- High-performance double-glazed windows offered to owners.
- Overall heating energy savings of 64 per cent.

Description

This large apartment building in central Slovenia was built in 1965. It contains 55 apartments that are owned by various private owners. The building was designed and built during the period when there was no regulation and no requirements regarding the thermal insulation and energy efficiency of buildings. Thus, the brick outer-wall structure had a U value of approximately 1.6W/m²K, and the windows (which were double glazed) had a U value of approximately 2.3W/m²K, with high air leakage.

Figure 5.30 The building before renovation

Source: EI-Education

Figure 5.31 The building after renovation

Source: EI-Education

Refurbishment Process

The reasons for the renovation were the building's poor maintenance and low-quality window frames; but a further intention of the housing association was to implement energy-saving measures and to improve the overall appearance of the building. Owners were given the option of window replacement; but only 40 per cent accepted this option.

Sustainability Improvements

Table 5.18 Sustainability improvements in Case Study 5.9

Walls	External insulation.	U value: $0.35W/m^2K$
Floors	No change.	
Roof	Insulation added.	U value: $0.4W/m^2K$
Windows	Double glazed with argon fill and low e coating; insulated frames (40% of owners only).	U value: $1.4W/m^2K$
External doors	Unknown.	
Other sustainability features	Unknown.	
Airtightness	Unknown.	
Ventilation	Improved airtightness around windows.	
Heating	No change.	
Renewable energy sources	None.	

Comfort and Overheating

The benefits for the occupants are reported as lower heating costs, higher levels of thermal comfort, improved aesthetics and an overall increase in the value of the building.

Change in Energy Use and CO_2 Emissions

Table 5.19 Change in energy use before and after refurbishment in Case Study 5.9

	Before refurbishment (calculated)	After refurbishment (calculated)	Savings
Space heating used	8693kWh/year	3166kWh/year	64% (5527kWh/year)

Costs

The project required what was considered a considerable investment, and the state subsidy for energy refurbishment was used to support the organization and execution of the works.

Planning Issues

None were reported.

Conclusions

The insulation measures have resulted in significant energy and costs savings; furthermore, users' increased awareness and changed living habits are expected to have additional positive effects on the savings.

The success of this project is claimed to be due to the cooperation of flat owners and users, as well as the high level of consensus achieved, which is necessary to carry out technical improvements to a building like this.

Case Study 5.10 Sweden: Alingsås

- 1970s apartment block fully renovated.
- 480mm of external insulation added to long walls.
- Heating from ventilation system, with heating coil added.
- Triple-glazed windows.
- Demonstration project for large estate refurbishment.

Description

Alingsås is a small but growing town near Gothenburg on the west coast of Sweden. The case study block of 18 apartments built during 1970 is owned by the local public housing company. The brick façade needed to be changed anyway and the additional expense for adding new and additional insulation was not perceived as being high, thus making the decision to renovate to *PassivHaus* house standards easier. The windows and the ventilation system also needed to be upgraded, which was a very good starting point for an energy-efficient renovation. The block is a concrete frame structure with gable ends and apartment-dividing transverse walls used as the load-bearing structure. This structure was the most common building system during the period of 1960 to 1975 in Sweden.

Refurbishment Process

A full study of the expectations and needs of the different tenants living in the building was performed concurrently with the planning process of the renovation. Initially, the renovation was planned to be performed from the outside so that each apartment could be renovated in a maximum of one week, making it possible for the tenants to stay in their apartments during the renovation. During the planning process, when the renovation assumed a greater scope, it was realized that it was necessary to evacuate the tenants during the building work; more information was thus provided to the tenants. An empty apartment was used as a full-scale demonstration apartment, where the tenants could meet and ask questions regarding the

Figure 5.32 Block before refurbishment

Source: Ulla Janson

Figure 5.33 Block after refurbishment

Source: Ulla Janson

renovation (e.g. focusing on the problems of air infiltration from joints in prefabricated elements, cold surfaces and low internal temperatures).

Sustainability Improvements

Table 5.20 Sustainability improvements in Case Study 5.10

Walls	Walls along the long sides: 480mm external insulation.	U value: 0.11W/m²K
	End walls: 130mm mineral wool (unchanged); new balconies founded on plinths and separate from the main building.	
Floors	Concrete slab floors: 40–60mm of cellular concrete added with 60mm expanded polystyrene insulation.	U value: 0.16W/m²K
	Foundation construction above ground: insulated with 270mm of expanded polystyrene.	
Roof	300mm of loose mineral wool insulation outside on the existing roof; a board of 100mm mineral wool was added to reduce overnight cooling by radiation in order to avoid moisture on the inside of the wooden roof construction.	U value: 0.13W/m²K
Windows	Triple-glazed windows; outer panes have low emissivity coating and gaps filled with krypton; frames are highly insulated.	Full window U value: 0.85W/m²K
External doors	New glazed entrance doors.	U value: 0.75W/m²K
Other sustainability features	Acoustic treatment between apartments. LED lighting in stairwells. Low-flow hot water taps.	
Airtightness	New wall construction and windows are carefully sealed.	Air change rate: 0.1ACH
Ventilation	Separate ventilation unit installed in each apartment, with heat exchanger.	Heat recovery: 80%; fan power: 2 x 58W
	Kitchen extract fans.	
Heating	Heating coil in the ventilation systems from original district heating connection.	
	An electrical towel rail of 70W installed in the bathrooms.	
Renewable energy sources	Solar thermal panels will be added to the district heating system later on.	

Comfort and Overheating

During the cold season, some apartments had low indoor temperatures. The main problems that were unresolved occurred on the ground floor, and there were varying underlying reasons in different apartments. One conclusion was that part of the cause was the temperature of the ground slab, and a revised construction was suggested for subsequent improvements.

Figure 5.34 Plinths for balconies cast separate from the building construction in order to avoid thermal bridges

Source: Ulla Janson

Figure 5.35 Thermal bridge insulation around the foundation slab construction

Source: Ulla Janson

Figure 5.36 Model of exterior long side wall

Source: Ulla Janson

The indoor temperature during the summer months was perceived to be very comfortable in the households living on the ground floor, according to the interviews with the tenants. However, the tenants living on the third storey reported very high temperatures and discomfort during summer; one tenant had installed Venetian blinds.

Change in Energy Use and CO_2 Emissions

Table 5.21 Change in energy use before and after refurbishment in Case Study 5.10

	Before refurbishment (average per apartment; measured data)	After refurbishment (average per apartment; measured data)	Savings
Space heating used	9136kWh/year	2119kWh/year	77% (7017kWh/year)
Electricity used	4675kWh/year	3465kWh/year	26% (1210kWh/year)
Domestic hot water	3313kWh/year	1282kWh/year	61% (2031kWh/year)

Costs

The budget for each apartment was set at 580,000 Swedish kroner for the higher standard, 270,000 kroner for the energy-saving measures and 140,000 kroner for maintenance, giving a total investment cost of 1 million kroner per apartment. The actual cost for renovating the demonstration building was higher than the amount allotted; but the subsequent costs for renovating the case study buildings largely fell within the budget.

The project manager believes that the energy-saving investment will be paid back within ten years, depending upon energy prices.

Conclusions

The client believes that the measured results of energy use after renovation are impressive and demonstrate a high-quality product delivered by the general contractor and their subcontractors. The results show that it is possible to renovate a building to achieve high energy effeciency using traditional building materials and employing regular contractors.

The client was very satisfied with the result of the demonstration project, and in August 2008 the planning process of phase two began, involving the renovation of two more buildings in the Brogården area.

Figure 5.37 Apartment buildings in the Brogården area, Alingsås

Source: Ulla Janson

Case Study 5.11 The Netherlands: Raamsdonk

- The housing association carried out a package of improvements to terraced housing.
- Careful consultation with tenants to manage the improvements.
- External insulation with insulated cladding bricks.
- Energy savings in the range of 50 per cent, on average, and a maximum of 70 per cent.

Description

This terrace of 42 houses in central Netherlands close to Breda, owned by the Volksbelang Housing Association, was built between 1963 and 1969. The tenants did not originally consider the proposed measures to be necessary and there had been no complaints. The housing association decided to give a guarantee to the tenants that they would not have much higher expenses because of the renovation. The renovation work was carried out between 2000 and 2002.

Refurbishment Process

The Volksbelang Housing Association's intention was threefold: improvement of the construction, improvement in quality of living and energy savings.

Energy-saving measures were to be installed during the overall planned renovation. In order to convince the tenants and set a good example, a model house was used in which all proposed measures were installed. At the same time as the energy measures, the tenants were offered other improvements in a form of packages, which included a dormer window in the bedroom, improved access to the attic and a more luxurious version of the kitchen units. For tenants requesting these items, the rent would be increased. In addition, storage rooms were renovated in all houses.

Figure 5.38 The houses before renovation

Source: Volksbelang Housing Association

Figure 5.39 The houses after renovation

Source: Volksbelang Housing Association

Sustainability Improvements

Table 5.22 Sustainability improvements in Case Study 5.11

Walls	External insulation, including insulated cladding bricks.	U value: 0.35W/m²K
Floors	No action undertaken.	
Roof	Insulation added to uninsulated roofs.	U value: 0.3W/m²K
Windows	High-efficiency double glazing.	U value: 1.3W/m²K (including frame)
External doors	Unknown.	
Other sustainability features	Chimneys insulated externally.	
Airtightness	Unknown.	
Ventilation	Mechanical ventilation with heat recovery.	Heat recovery: 80%
Heating	Gas-fired individual boilers, some upgraded.	
Renewable energy sources	None.	

Change in Energy Use and CO$_2$ Emissions

Table 5.23 Change in energy use before and after refurbishment in Case Study 5.11

	Before refurbishment (monitored figures)	After refurbishment (monitored figures; best average)	Savings
Gas used	19,200kWh/year	5760–9600kWh/year	50–70% (9600–13,340kWh/year)

Costs

The total cost for the renovations and improvements averaged 39,900 Euros per dwelling, of which 3300 Euros came from a government subsidy for energy-saving measures. The Volksbelang Housing Association set the investment costs to minimize the increase in rent. The investment is likely to be paid back within several years. A guarantee was given to the tenants that during the first five years, the package of rent, energy costs and maintenance would not increase by more than 0.5 per cent compared to the housing association's average rent.

Planning Issues

None were reported.

Conclusions

Energy savings have been achieved by means of proved and efficient technology, and are therefore seen as feasible for most housing associations. The external insulation resulted in fewer disturbances for tenants, higher efficiency and lower fuel costs; in addition, the houses now have a fresh appearance.

The housing association is very satisfied with the result and intends to include energy-saving measures in future renovation projects – for example, in 2006 the Volksbelang Housing Association started a renovation project including a further 95 houses based on the principles and experience gained from the renovation of the dwellings in this project.

6 Practical Details and Choosing the Best

This book does not provide detailed information on the practicalities of sustainable refurbishment – there are many publications that already deal with this which should be used for detailed design and specifications. This chapter outlines each building element, discusses the options related to different conditions, and references the most useful guides available and where detailed information can be found. In most situations, a range of solutions can be used successfully, depending on the standards aimed at, the conditions found, the budget available, planning and visual restrictions, and social issues such as acceptable levels of household disruption.

Guidance and experience has been built up over many years in all countries and detailed information should be obtained from national bodies specializing in energy use and sustainable housing. National websites include:

- www.energysavingtrust.org.uk/business/ Business/Resources/Publications-and-Case-Studies in the UK;
- www.dena.de in Germany; and
- http://www2.ademe.fr in France.

Much UK information is contained in Energy Saving Trust publications, which have been developed from UK government research and development, field trials, monitoring programmes and experience gleaned from housing managers, owners and activists. There are several guides covering general refurbishment:

- *Sustainable Refurbishment CE309 2010: Towards an 80% Reduction in CO₂ Emissions, Water*

Efficiency, Waste Reduction, and Climate Change Adaptation (Energy Saving Trust, 2010);
- *Energy Efficient Refurbishment of Existing Housing* (Energy Saving Trust, 2007);
- *Practical Refurbishment of Solid Walled Houses* (Energy Saving Trust, 2006);
- *Energy Efficient Refurbishment of Non-Traditional Houses* (Energy Saving Trust, 2006);
- *Refurbishing High Rise Dwellings: A Summary of Best Practice* (Energy Saving Trust, 2006);
- *Energy Efficient Domestic Extensions* (Energy Saving Trust, 2005);
- *Energy Efficient Historic Homes: Case Studies* (Energy Saving Trust, 2005);
- *Energy Efficient Loft Extensions* (Energy Saving Trust, 2005).

Energy Saving Trust guides covering specific areas of energy refurbishment are listed in each section below.

Improving the Insulation of all External Elements

Minimizing heat loss through the building envelope is the basic aspect of sustainable refurbishment, and all elements require attention before issues of heating supply and use are dealt with.

Insulation types and standards

The insulation values (and costs) of insulation materials vary considerably and the choice is likely to depend upon available space or acceptable loss of space. The additional cost of thicker insulation is likely to be a small part of the overall refurbishment cost; thus, when refurbishment work is planned, the maximum thickness possible according to the circumstances will provide the most cost-effective long-term benefits.

Table 6.1 shows the thermal conductivity of common generic insulation materials used in houses: the lower the conductivity, the better the insulation value. Thus, the better-insulating materials can give U values twice as high as others.

Table 6.1 Thermal conductivity of common insulation materials used in housing

Insulation type	Typical thermal conductivity (W/mK)
Aerogel	0.013–0.018
Phenolic foam	0.022
Polyisocyanurate and polyurethane	0.023
Mineral wool (slab)	0.035
Expanded polystyrene	0.038
Cellulose fibre (recycled)	0.040
Wood fibre	0.044
Sheep's wool	0.04–0.05

Two standards are used in this book to give targets for insulation levels: the UK Energy Saving Trust's best practice standards and the German *PassivHaus* standards. The *PassivHaus* standards are generally considered to be the ultimate target, producing housing that requires very little or no space heating. There are numerous practical reasons for adopting lower standards; but best practice standards are recommended as the minimum on the basis that refurbishment is only carried out infrequently and higher insulation standards will always give permanent and long-term benefits.

Figure 6.1 The Energy Saving Trust

Insulation: Environmental considerations

As well as choosing insulation materials on the basis of thermal conductivity, cost and thickness, other environmental aspects can be considered:

- Natural materials such as cellulose fibre from recycled paper, wood fibre and sheep's wool have a very low environmental impact and may be preferred to manufactured and fossil-fuel-based materials (such as polystyrene).
- Embodied energy will be lower in natural materials compared with manufactured products (e.g. mineral fibre).
- Health-and-safety issues include fibres in mineral and glass wool (although there is no evidence of this); off-gassing related to indoor air quality; and the production of poisonous gases in the event of fire.
- Longevity issues relate to cellulose insulation getting wet and becoming compacted, and animal attacks on natural materials if not properly treated.

Insulating walls

UK best-practice refurbishment requires a wall U value of 0.3W/m²K, which indicates an insulation thickness of between 80mm and 120mm depending upon the existing wall construction and the insulation type used. In order to achieve *PassivHaus* standards, wall insulation thicknesses of 200mm and above are normally required.

Solid walls

The most common walls in existing housing are solid. External insulation has the advantages of less disruption inside of the house, retaining thermal mass inside, dealing with potential cold bridges, and weather-proofing, but has the disadvantages of higher cost, the need to extend roofs and window sills, and the necessity of moving drainage pipework. Changing the appearance of the building can be either positive if the existing finish is poor, or negative if existing architectural features are covered up. Internal insulation has the advantage of being less expensive and not changing external appearance, but the disadvantage of internal disruption, loss of internal space, the need to move power points, skirting boards, radiators, wall fittings, etc., difficulty in fixing heavy items to the wall, the need to replace ceiling mouldings, and dealing with cold bridging. The use of Aerogel, the highest-value insulation, is becoming more common where very thin insulation is required and can be used to avoid replacing mouldings and losing room space.

External insulation

There are two common methods of external insulation: wet render and dry cladding.

Wet render systems consist of:

- insulant;
- adhesive mortar and/or mechanical fixings;
- profiles and edgings used on corners, at damp-proof course level, window reveals, verges and copings;
- a base-coat render, incorporating a glass fibre, plastic or metal mesh;
- a top-coat render, with or without a finish.

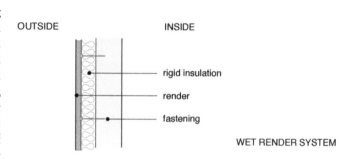

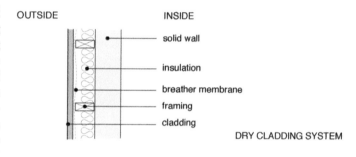

Figure 6.2 External wall insulation

Dry cladding systems consist of:

- insulant;
- a supporting framework or cladding fixing system fixed to the wall;
- a ventilated cavity;
- cladding material and fixings, timber panels, stone or clay tiles, or brick slips.

Wet render systems are likely to be cheaper, but dry cladding systems can be useful where planning permission is an issue and existing appearances need to be maintained. Several issues must be considered, including fire barriers in multi-storey buildings; the strength of the finishing; the need to provide reduced thicknesses of insulation in window and door reveals to stop cold bridging; and sealing of joints around the insulation to stop water ingress.

The UK Insulated Render and Cladding Association (INCA) has a list of proven systems and approved installers. The work is best carried out by a specialist installer.

Reference: *External Insulation Systems for Walls of Dwellings (CE118/GPG293)* (Energy Saving Trust, 2006)

Internal insulation

There are three common methods of internal insulation: laminated insulation board fixed directly to the wall; rigid insulation between battens fixed to the wall; and a frame with insulation leaving an air gap between insulation and the wall. Other proprietary systems are also available.

Directly applied internal insulation consists of:

- plasterboard laminated to insulation board;
- a built-in vapour-control layer to stop moist internal air from condensing on the cold wall;
- fixed to the wall with adhesive, plaster dabs and/or mechanical means;
- continuous ribbons of plaster adhesive at the wall perimeter and around all openings (such as sockets and plumbing) to prevent cold air behind the insulation from leaking into the house.

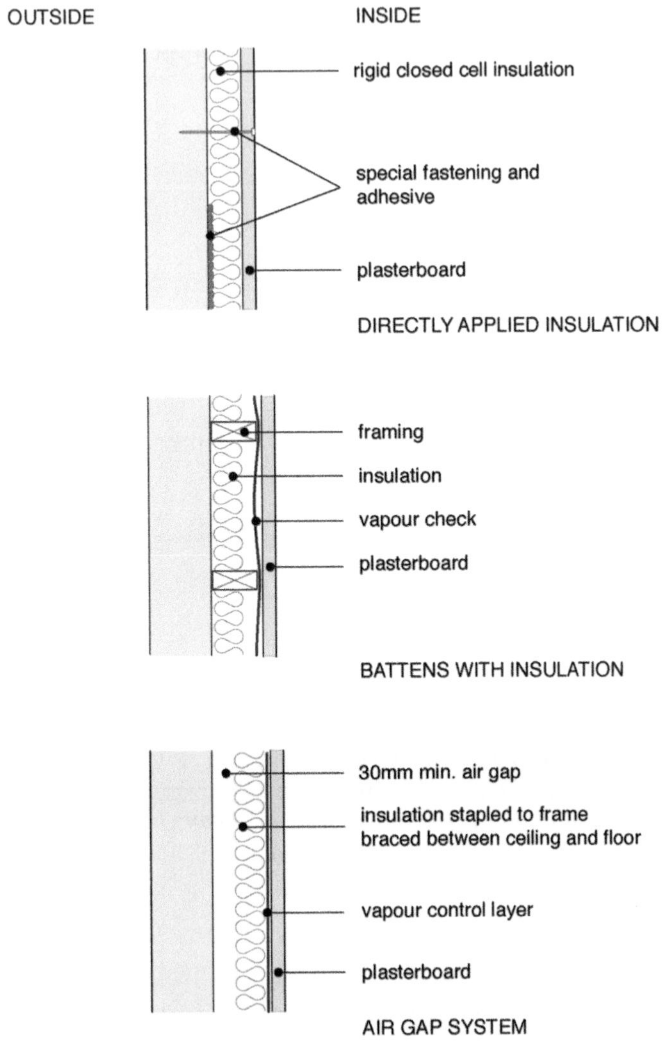

Batten with insulation systems consist of:

- vertical timber battens or metal furrings fixed to the wall;
- rigid or semi-rigid insulation boards between

OUTSIDE   INSIDE

rigid closed cell insulation

special fastening and adhesive

plasterboard

DIRECTLY APPLIED INSULATION

framing

insulation

vapour check

plasterboard

BATTENS WITH INSULATION

30mm min. air gap

insulation stapled to frame braced between ceiling and floor

vapour control layer

plasterboard

AIR GAP SYSTEM

Figure 6.3 Internal wall insulation

the battens or flanged paper-faced quilt insulation, with joints taped, stapled to the battens;
- a continuous vapour-control layer;
- plasterboard.

Air-gap systems consist of:

- timber or metal frame braced between the floor and ceiling and kept clear of the external wall;
- insulation stapled to the frame leaving a 30mm air gap between insulation and wall;
- a continuous vapour-control layer;
- plasterboard.

Batten and air-gap systems are useful where the existing wall is uneven or has a history of damp penetration. Where services penetrate the plasterboard and vapour barrier, adequate airtightness must be achieved to prevent cold air and moisture leakage. Insulation should be returned into window and door reveals to stop condensation on these potential cold bridges.

References: *Internal Wall Insulation in Existing Houses* (Energy Saving Trust, 2008); *Practical Refurbishment of Solid Walled Houses* (Energy Saving Trust, 2006)

Cavity walls

The majority of existing brick and block cavity walls up to 12m in height are suitable for cavity fill. Some products are independently certified for use in walls up to 25m. Suitability depends mainly on local exposure to driving rain and the condition of the existing construction. Cavity walls which are partially filled with insulation fixed to the internal leaf- and timber-framed buildings are not suitable for retrospective cavity fill.

The cavity fill process consists of:

- drilling injection holes through mortar joints;
- installing cavity barriers to prevent the fill from entering the cavities of adjacent properties;
- sleeving (or sealing and replacing, if obsolete) air ventilators that cross the cavity;
- injecting the fill material into the wall cavity and carrying out quality checks;
- sealing the injection holes with colour-matching mortar or render.

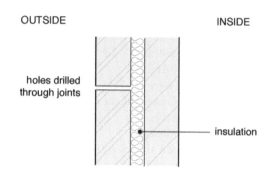

OUTSIDE   INSIDE

holes drilled through joints

insulation

Figure 6.4 Cavity wall insulation

Cavity wall insulation will normally only provide around 50mm of insulation; but additional insulation can be provided internally or externally, as described above. External wall insulation may not be suitable for cavity walls with a clear cavity (e.g. those with partial fill) as convection of the air in the cavity may reduce the thermal performance of the wall.

Reference: *Cavity Wall Insulation in Existing Dwellings: A Guide for Specifiers and Advisors* (Energy Saving Trust, 2007)

Insulating ground and exposed floors

The depth of insulation that can be used to insulate ground floors will often depend upon the space available. UK best-practice refurbishment requires a floor U value of $0.23W/m^2K$, which indicates an insulation thickness of between 125mm and 175mm depending upon the existing construction and insulation type. In order to achieve *PassivHaus* standards, floor insulation thicknesses of 250mm are required.

Suspended timber floors

There are two common ways to insulate a suspended timber floor: lifting the existing floorboards and laying insulation between the joists, or fixing insulation from below if there is access (e.g. from a cellar).

Insulation from above consists of:

- lifting the existing floorboards (and skirting boards);
- fixing netting over the joists;
- filling the full depth of the joists with insulation;
- replacing floorboards;
- sealing between the boards and below skirting and service entry points.

If the existing floor joists are not in good condition, consideration should be given to replacing the timber floor with an insulated concrete floor (see below).

Insulation from below consists of:

- sealing joints between boards and below skirting and service entry points from below;
- fixing insulation between and below the joists;
- fixing plasterboard to the underside of the joists (to provide fire resistance); or
- fixing rigid insulation underneath an existing basement ceiling.

This also applies to suspended upper floors, such as those with rooms above garages, walkways and recesses.

Concrete floors

Existing concrete floors can be insulated with insulation from above, and new concrete floors can be insulated above or below.

Insulation of existing ground floors consists of:

- rigid insulation on top of the slab;
- timber battens at thresholds with metal nosing;
- a vapour-control layer on the insulation;
- chipboard flooring;

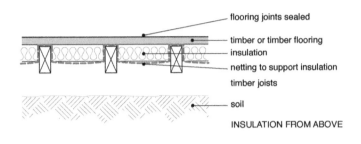

flooring joints sealed
timber or timber flooring
insulation
netting to support insulation
timber joists
soil
INSULATION FROM ABOVE

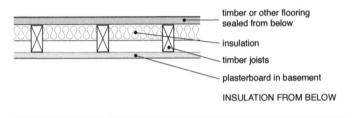

timber or other flooring sealed from below
insulation
timber joists
plasterboard in basement
INSULATION FROM BELOW

Figure 6.5 Suspended timber floor insulation

- floor finish;
- refixing skirting boards and reducing door heights.

Insulation above new concrete floors consists of:

- a damp-proof membrane on the slab linked to the damp-proof course (DPC) in the wall;
- rigid insulation on top of the slab;
- chipboard flooring;
- floor finish.

Insulation below new concrete floors consists of (from the bottom):

- sand bedding on the existing ground;
- damp-proof membrane to protect insulation (if necessary);
- rigid insulation with an upstand around the perimeter;
- concrete floor slab;
- damp-proof membrane linked to the DPC in the wall;

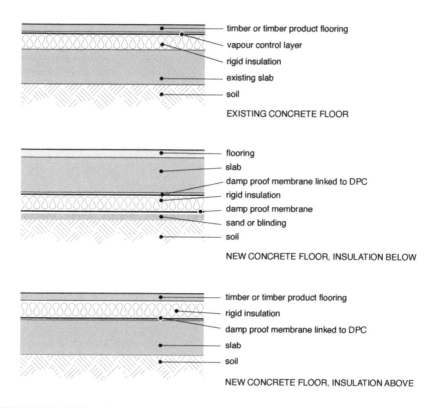

Figure 6.6 Concrete floor insulation

- screed or chipboard flooring;
- floor finish.

Reference: *Practical Refurbishment of Solid-Walled Houses (CE184)* (Energy Saving Trust, 2006)

Insulating roofs and exposed ceilings

UK best-practice refurbishment requires a roof U value of $0.16W/m^2K$, which indicates a total insulation thickness of between 250mm and 300mm depending upon the insulation type. In order to achieve *PassivHaus* standards, roof insulation thicknesses of between 300mm and 400mm are required.

Laying insulation flat on a ceiling in an open loft is the easiest method of insulating; but sloping roofs can be insulated at rafter level, and flat roofs above or below, although these methods require special attention.

Insulation at ceiling level

Insulation in lofts consists of (from the bottom):

- eaves ventilators fitted to stop insulation blocking ventilation paths;
- insulation between joists;
- additional insulation laid across joists;
- electric cables lifted above insulation;
- insulation omitted beneath water tanks to stop freezing.

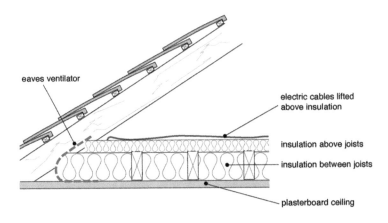

Figure 6.7 Ceiling-level roof insulation

Gaps around light fittings and opening into the loft space should be sealed to reduce air and moisture movement. If an area of storage in the loft is required, it is best constructed using insulation-backed decking above the insulated joists.

Insulation at rafter level

If the roof is being replaced, insulation can be fixed above and between the rafters, raising the roof slightly; if not, the insulation can be fixed between and below, reducing internal space.

Insulation above the rafters consists of (from the top):

*   an air gap between the roofing and insulation of 50mm;
*   rigid insulation on top of the rafters;
*   insulation in between the rafters;
*   a vapour barrier;
*   plasterboard across the inside rafters.

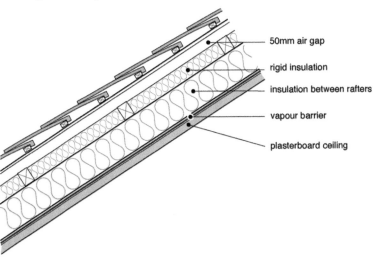

Figure 6.8 Rafter-level roof insulation

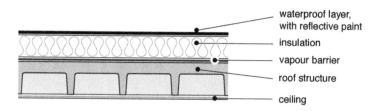

Figure 6.9 Flat roof insulation (a warm roof system)

If there is a room in the roof space, its other areas will also need insulating. The lower stud walls can be insulated like the rafters, while any gable or party wall should also be insulated if the adjacent properties do not also have rooms in the roof voids. These can be considered in the same way as solid walls. The ceiling above the room in the roof can be treated in the same way as either a normal insulated ceiling, or if there is insufficient space, then insulation can be placed between and below the ceiling joists.

Insulation of flat roofs

Concrete and timber flat roofs are best treated with insulation above the roof structure, known as a warm roof system, in order to avoid interstitial condensation.

Where the roofing is being replaced, the warm roof system consists of (from the top):

- a waterproof layer;
- rigid insulation;
- a vapour check;
- structural roofing.

Where the waterproof roofing is not being replaced, insulation can be placed above the existing roofing; the 'inverted warm deck' system consists of (from the top):

- a ballast layer to hold insulation down;
- rigid waterproof insulation;
- the existing waterproof layer;
- structural roofing.

As rain will always penetrate the insulation and reduce its insulation value, the insulation should be of a type that is unaffected by moisture. Additional insulation will be needed to achieve

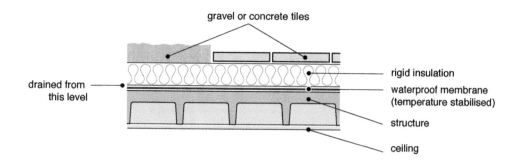

Figure 6.10 Flat roof insulation (an inverted warm roof system)

the same thermal performance as a traditional warm deck construction.

Where it is not possible to construct a warm deck, internal insulation can be used. This 'cold deck' system is susceptible to internal condensation on the cold deck surface and thus requires a highly effective internal vapour barrier, with complete sealing at wall junctions and around any service penetrations. Adequate ventilation under the structure is essential to avoid a build-up of moisture.

A cold deck system consists of:

- a waterproof layer;
- structural roofing;
- a ventilated space;
- insulation mechanically fixed to the structure;
- a vapour barrier;
- ceiling plasterboard mechanically fixed to the structure.

Insulation-backed plasterboard, with a built-in vapour barrier, is commonly used. Joints must be sealed to maintain a continuous vapour barrier.

Reference: *Practical Refurbishment of Solid-Walled Houses (CE184)* (Energy Saving Trust, 2006)

Insulated windows and doors

UK best-practice refurbishment requires a window U value of 1.5W/m²K for the complete window plus frame, which indicates a double-glazed window with low e coating and argon fill, as well as an insulating frame and spacer between the layers of glass. In order to achieve *PassivHaus* standards, a U value of 0.85W/m²K is required, indicating triple glazing. In the UK, a window's energy rating is assessed according to the British Fenestration Rating Council's (BFRC's) system, which combines the U value of the glazing, the U value of the frame, the proportion of the frame in the window, the solar transmittance of the glazing (its g value), and the airtightness of the unit as a whole. This system, set up in conjunction with European partners, compares the overall energy performance of windows based on the total annual energy flow through the window. Best practice requires a BFRC rating in band C or better for windows, patio doors and French doors.

Existing windows are best replaced by new frames with double glazing and complete with draught-stripping. New double glazing can sometimes be fitted into existing frames depending upon the type, condition and size; but any work must include installing effective draught-stripping, using neoprene strip routed into the timber frames. Double glazing with a 4mm gap instead of the normal 12mm gap may be useful where space is restricted, and can be used, in some instances, to satisfy planning restrictions.

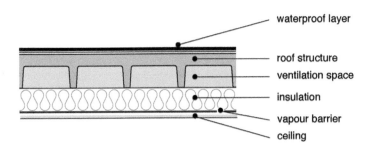

waterproof layer

roof structure

ventilation space

insulation

vapour barrier

ceiling

Figure 6.11 Flat roof insulation (a cold roof system)

Figure 6.12 Double-glazed sash window
with spring balance

Source: Simon Burton

Reference: *Windows for New and Existing Houses* (Energy Saving Trust, 2006)

Avoiding cold bridging at junctions and balconies

Cold bridging occurs where insulation is missing or where solid materials span between the outside and inside. Avoidance of cold bridges is important in planning, system choice, and design and implementation in order to reduce heat loss and avoid localized condensation and mould growth. Cold bridging commonly occurs at the junctions of walls, floors and roofs, and also around windows and doors, as well as via protruding balconies. Most design guides cover the subject of avoiding cold bridging, and *PassivHaus* literature provides detailed information for new-build situations, some of which is applicable to refurbishment.

External insulation, with insulation of window and door reveals, avoids most cold bridges except at ground-floor level and protrusions such as balconies and roof upstand walls. Special attention is required to address this.

Internal insulation requires thinner insulation in window and door reveals, and returning insulation, where possible, along internal walls.

Vacuum glazing, comprising two panes of glass sealed together and separated by glass beads in a grid pattern, is now commercially available (Pilkington 'EnergiKare Legacy'). The gap is around 2mm and the space is evacuated and sealed with a cap. Vacuum-glazed windows can achieve whole window and window frame U values of $1.8W/m^2K$ with a very thin profile.

Doors, whether unglazed or partially glazed, can have insulated cores between the two outer surfaces to achieve U values of $0.8W/m^2K$.

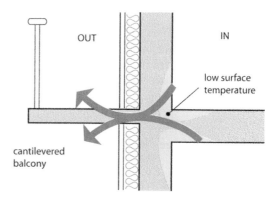

Figure 6.13 A cold bridge left by external insulation

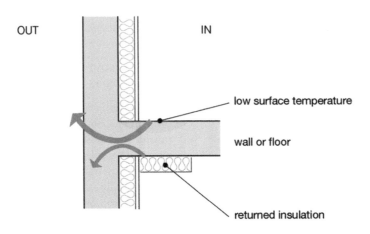

OUT   IN

low surface temperature

wall or floor

returned insulation

Figure 6.14 A cold bridge left by internal insulation

Window and door frames are also a potential cold bridge and where replaced should be thermally broken or feature low thermal conductivity.

Reference: *AECB CarbonLite Programme: Delivering Buildings with Excellent Energy and CO2 Performance. Volume 5: Steps Two and Three Design Guidance, PassivHaus/Gold Standard* (AECB, 2009)

Ventilation that Is Adequate and Efficient

The numerous cracks and air leakage paths in the external structure need to be sealed, opening windows and doors draught-stripped, and controllable active or passive ventilation provided to supply the necessary fresh air. All heating appliances should be 'room sealed'; but open coal- or wood-effect fires, as well as some biomass stoves, will need specific ventilation paths.

External insulation can be very effective in reducing air leakage.

Improving the airtightness of the structure

All of the following actions will reduce air leakage:

- Repair damage to mortar joints and fill any holes in the external walls.
- Apply sealant materials to fill the gaps around windows and doors.
- Apply an external mastic seal to all window and door frames.
- Apply a bead of mastic to seal any internal gaps between the wall reveals/window boards and windows and external doors.
- Seal gaps around service pipes and the cables that pass through external walls, ceilings and ground floors, especially behind fittings, units and boxing out.
- Apply a heatproof seal around boiler flue pipes where they pass through the external wall/ceiling.
- Ensure that the loft hatch fits into the frame and apply draught-stripping between the hatch and the frame.
- Draught-proof suspended timber floors when they are being insulated, or if not by laying hardboard sheeting over the top (to cover gaps between the floorboards), sealing around the edges of the rooms and gaps around service pipes.
- In dry-lining applications, inject continuous ribbons of expanding polyurethane foam

between the edges of the plasterboard and insulation sheets and the structural wall.

- Block off existing unused chimneys; although a grille to enable ventilation may be necessary, this is best done from outside if possible.

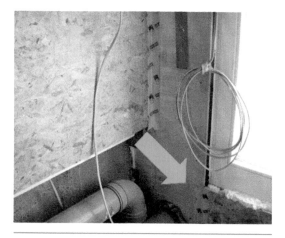

Figure 6.15 Sealing the air leak

Source: Paul Jennings

The success or otherwise of measures to reduce air leakage is best tested with blower door equipment, which will show the air changes per hour (ACH) at a standards pressure of 50 pascals (Pa). The *PassivHaus* standard proposed for refurbishment is 1ACH at 50Pa.

References: *Improving Air-Tightness in Dwellings* (Energy Saving Trust, 2005); *Sustainable Refurbishment: Towards an 80% Reduction in CO_2 Emissions, Water Efficiency, Waste Reduction, and Climate Change Adaptation* (Energy Saving Trust, 2010)

Draught-stripping windows and doors

Replacement doors and windows will come with built-in draught-stripping; but where existing units or frames are retained, the openings can be retrofitted with draught-stripping. Timber frames and doors in unwarped good condition, including both hinged and sash windows, can be

grooved to receive draught-stripping to all edges. Different types of draught-stripping are appropriate for different situations, and these include brush and compression strips. Door and French window thresholds can be fitted with spring-loaded draught-stripping units.

Providing adequate and controllable ventilation

Either passive or whole-house mechanical ventilation systems can be used in existing housing. Passive ventilation may be easier to use in most dwellings; but for the high overall efficiency needed to achieve *PassivHaus* standards, whole-house mechanical ventilation with heat recovery is required.

Passive ventilation

Passive ventilation requires air-intake background or trickle-ventilator units above windows with passive-stack extract systems or mechanical extract fans in kitchens and bathrooms in order to remove pollutants and provide fresh air.

Low-power extract fans are available and they are best controlled by humidistats. Heat-recovery extract and supply fans for one room are available. A passive-stack system can be installed in some dwellings, consisting of a vertical, or near vertical, series of ducts that connect kitchens and bathrooms directly with the outside through the roof. Moist air is extracted by the stack effect, as well as by the effect of the wind blowing over the roof.

Reference: *Sustainable Refurbishment: Towards an 80% Reduction in CO_2 Emissions, Water Efficiency, Waste Reduction, and Climate Change Adaptation* (Energy Saving Trust, 2010)

Whole-house mechanical ventilation

Ducting built into the dwelling, together with a fan, extracts moist warm air from kitchens and bathrooms and draws in replacement air, which is preheated by a heat exchanger on the outgoing air and ducted to other rooms. No background ventilation is required; in fact, the dwelling must be made very airtight to ensure that the system works efficiently.

As long as the final air leakage rate is less than $5m^3/m^2$/hour at 50Pa, then significant energy efficiencies are possible with this type of system if it has a specific fan power of 1W/L/second or less and a heat recovery efficiency of 85 per cent or more. Considerable work will be necessary to improve the airtightness of the dwelling, as described above, to achieve this standard.

Figure 6.16 Whole-house mechanical ventilation system

Source: DENA

Fitting the ductwork into an existing house may be difficult, depending upon space and the level of renovation being carried out.

Reference: *Sustainable Refurbishment: Towards an 80% Reduction in CO₂ Emissions, Water Efficiency, Waste Reduction, and Climate Change Adaptation* (Energy Saving Trust, 2010)

Increasing Solar Gain

Sun entering a dwelling through east, south and west windows, as well as roof lights, assisted by thermal storage in floors and other thermal mass, can displace fossil-fuel heating. As with daylighting, adding south-facing windows in an east–west-facing house can provide useful solar gain, and this can be optimized by a heat recovery ventilation system which will distribute the heat around the house.

Efficient Space Heating

Some space heating is usually needed in all housing in order to warm up from a cold start, even if solar and internal gains can maintain adequate temperatures in dwellings refurbished to *PassivHaus* standards. Dwellings sustainably refurbished even to UK best-practice standards will require only small space heating inputs, and this should be calculated based on the new insulation standards to avoid boiler oversizing. A range of heating methods and appliances may be appropriate depending upon the availability of a local communal heating supply, gas supply mains or biomass resource.

Heat recovery from extract air can provide a large proportion of space heating needs, but is only applicable where a whole-house ventilation system is used. Individual room extract and supply units only provide heating on a localized basis, and this may not be needed when available (e.g. in kitchens when cooking is taking place).

Under-floor heating can be used to make lower air temperatures acceptable, enabling lower boiler flow and return temperatures to generate greater efficiency via more time spent in condensation mode. Heat pump systems are also more efficient when supplying heat at lower temperatures.

Figure 6.17 Under-floor heating system

Source: Nick Rouse

Reference: *Whole House Boiler Sizing Method for Houses and Flats* (Energy Saving Trust, 2003)

Linking to a communal heating scheme

Where a communal block or district heating system is available, this is likely to provide the best space (and water heating) solution, assuming that waste heat, biomass or combined heat and power (CHP) are in use. Individual-dwelling heat metering and good internal controls, as described below, are necessary for efficient operation and use.

Reference: *The Applicability of District Heating for New Dwellings* (Energy Saving Trust, 2008)

Gas and oil boilers, including controls

Where gas is available, a modern room-sealed modulating gas condensing boiler will provide efficient space heating. The boiler will need to be sized for warm-up from the cold at the external design temperature; but if a combination boiler is used, this will be sized for hot water production and thus quite possibly oversized for space heating needs in a well-insulated house. A full package of controls is necessary (time clock, room thermostat and thermostatic radiator valves

on all radiators except in the room with the room thermostat), and it is important that the room thermostat is located and set so that it will turn off the boiler when the house is up to the required temperature. Sophisticated controls, such as optimal start or 'time proportional controllers', are likely to be unnecessary in a low-energy house. Radiators or under-floor heating can be used. Smaller and/or fewer radiators than in the pre-refurbishment dwelling will be necessary and the system bypass must be correctly fitted and controlled by a pressure-release valve. Oil-fired boilers may be appropriate where a mains gas supply is not available, and the same principles apply.

Reference: *Domestic Heating by Gas Boiler Systems: Guidance for Installers and Specifiers.* (Energy Saving Trust, 2008)

Heat pumps

Air-sourced heat pumps can provide low-carbon space (and water) heating in low-energy housing, particularly where solar water heating is fitted. Those with high annual coefficients of performance and controls that enable a temperature boost when providing domestic hot water will give the best performance. Heat pumps that use the house extract air as the heating source can be very effective and efficient. In a low heat-demand dwelling, the expense of providing a ground-sourcing heat pump system, though more efficient, is unlikely to be viable unless it is also used for providing cooling where this is necessary in hot climates. In houses with higher heat demands located off the mains gas supply, ground-sourced heat pump systems will be more efficient than those that are air sourced.

Biomass systems

Biomass heating can provide a low-carbon heat supply either as a stand-alone room heater or as a central heating boiler if a biomass supply is available locally and fuel storage is possible. Room-heating stoves using pellet or log fuel may be appropriate as the only space-heating system for small dwellings with low heat demand, and they are available with back boilers to provide domestic hot water. For larger dwellings with higher heat demand, biomass boilers with a wet heating system, using pellet, chip or log fuel, will be necessary. Most biomass burners take combustion air from the room and this will affect their location and ventilation requirements. The requirements for fuelling and ash removal must also be considered.

Figure 6.18 Biomass gasifier boiler

Source: Atmos

MicroCHP systems

Individual micro-combined heat and power (microCHP) systems using internal combustion or Stirling engines will not be appropriate for dwellings with small heat demand, and may only be efficient in larger dwellings where space

heating is required for long periods of time (e.g. where insulation levels are poor). Fuel cell-based CHP systems may become applicable in the future.

Reference: *Sustainable Refurbishment: Towards an 80% Reduction in CO$_2$ Emissions, Water Efficiency, Waste Reduction, and Climate Change Adaptation* (Energy Saving Trust, 2010)

Providing Domestic Hot Water (DHW) by Efficient Means

In a sustainably refurbished dwelling, domestic water heating can constitute the larger part of the heating demand. It is not only the generation of hot water that is important for efficiency, but also storage, delivery and usage. Solar systems supplying around 50 per cent of annual demand, and demand reduction by installing low-flow equipment, are important components in providing an efficient low-carbon domestic hot water (DHW) system.

If an existing central heating system is to be used and fitted with a new boiler, it is important to check that the system is designed to enable good water flow, is not blocked by sludge and has a bypass with a pressure release valve. Thermostatic radiator valves (TRVs) can and should be retrofitted if they don't exist.

Solar water systems

Solar collector panels can be retrofitted to any dwelling with south- or even east- and west-facing roofs, using either flat plate or evacuated tube systems. Evacuated tubes give higher performance but are more expensive. Space for hot water storage is necessary, sized according to the collector size and number of dwelling occupants. Systems must have drain-back components or use antifreeze in areas with sub-zero temperatures in winter to avoid freezing. Different systems are available using single- or double-storage cylinders. For providing top-up heating, not all combination boilers are compatible with solar systems as they may not be able to

Figure 6.19 Evacuated tube solar water heating

Source: Charlie Baker

take already hot water from a storage cylinder. In some countries, such as Greece and Cyprus, stand–alone systems incorporating collectors and storage are used, mounted on rooftops. Such systems are not used in many countries for visual and planning reasons.

References: *Solar Water Heating Systems: Guidance for Professionals, Conventional Indirect Models* (Energy Saving Trust, 2006); *Factsheet 3: Solar Water Heating* (Energy Saving Trust, 2006)

Linking to a communal heating scheme

Where a communal block or district heating system is available, this is likely to provide the best domestic hot water supply (see 'Efficient space heating'), in combination with a storage cylinder. Combining solar and district heating supplies may be possible and efficient if the district heating is not operational during summer, when the solar system will supply the majority of the heating demand.

Gas and oil boilers

Where gas is available, a modern room-sealed gas condensing boiler will provide efficient domestic hot water. This can be either a conventional boiler with a storage cylinder or a combination boiler, since the lower efficiency of a combination boiler in hot water mode will be offset by losses from the storage cylinder and pipework in a conventional system. Either can be equally efficient and the choice is likely to be made on the basis of the available space. Conventional boilers in conjunction with a storage cylinder are frequently used where solar water heating is to be employed; but some combination boilers can be used to top up solar hot water stored in a solar cylinder.

Heat pumps

Air-sourced heat pumps can be used to provide domestic hot water in conjunction with space heating. Those with high annual coefficients of performance and controls that enable a temperature boost when providing domestic hot water will give the best performance.

Direct electric heating

Direct electric heating for domestic hot water will not be a low-carbon option unless the electricity comes from a renewable resource, or most of the heating is solar and the backup only used occasionally, possible in very sunny southern areas. For isolated washbasins and showers, an instantaneous electric water heater may be the most efficient approach to reduce waste from long pipe runs.

Storage cylinders

In conventional systems, the hot water storage cylinder should ideally be located close to both the boiler and the bathroom and kitchen to reduce heat loss from pipes. Pipes between the boiler and cylinder should be insulated. The cylinder should have a high-efficiency heat exchange coil (or rapid recovery coil) and be factory insulated, ideally with an additional insulation jacket added *in situ* covering all the cylinder attachments that lose heat. A securely fixed and visible thermostat should be used.

Reducing hot water use

Showers with low-flow heads can use much less water than baths; but pumped 'power showers'

should be avoided as they can use a lot more hot water. Where taps are used for hand washing, reduced flow or spray taps will reduce water use, but are not appropriate for filling basins and sinks, or where a combination boiler is used as they are likely to have a flow too small to trigger the firing of the boiler.

Avoiding Overheating that Could Require Active Cooling

The issue of overheating needs to be considered for all dwellings being sustainably renovated in order to avoid the necessity of active cooling due to the effects of insulation and ventilation changes on heat gains and losses, and the usable thermal mass of the dwelling. Climate change is also likely to increase overheating in the future. Avoiding overheating will require three components: minimization of heat gains, together with the use of thermal mass and good controllable ventilation.

Internal heat gains

Insulation of domestic hot water pipework and storage cylinders will reduce heat gains to the dwelling, as will specification of efficient electrical equipment, including lighting, refrigerators and televisions.

External heat gains

Solar gain entering the house through windows is the common cause of overheating and can be reduced by providing external shading over south- and west-facing windows, planned to cut out sun during summer. East-facing windows will allow morning sun to warm the dwelling, which is likely to be beneficial and will not normally cause serious overheating. Horizontal shading is effective on south windows; but vertical shading is preferred on west windows as sufficient horizontal shading may reduce daylighting. Moveable external shading is more complex, but is effective in providing solar gain and additional daylighting when required.

Figure 6.20 Planting will provide solar shading

Source: Jes Mainwaring

Insulation of roofs is important in reducing heat gain through the structure, and treatment of flat roofs with reflective paint or using green/brown roofs is also effective.

Planting and vegetation

Trees can provide shading to the lower floors of a dwelling, and replacing hard surfaces by planting around the dwelling can lower external temperatures, thus reducing the temperature of air entering the house. This can help to counter the heat island effect in dense urban areas.

Internal thermal mass

Many older houses are built with a substantial internal mass in floors and external and partition walls. As long as this is not internally insulated, it can provide a coolth store and buffer against summer overheating. Using internal thermal mass is very important in reducing overheating and can easily be used in conjunction with good ventilation practices. Even if external walls are internally insulated, thus reducing available thermal mass, internal walls and solid floors and ceilings can provide sufficient mass. Burglarproof windows that allow night-time opening, preferably on both sides of the building to ventilate out heat built up during the day, will normally cool the building sufficiently before the next day. If cross-ventilation is not possible, a combination of high- and low-opening windows will encourage reasonable night ventilation.

Ventilation for cooling

Controlled ventilation during the daytime is also important to manage air movement depending upon whether the outdoor or indoor air is cooler. The designer should provide opening windows with variable openings at high and low levels, as well as windows that enable cross-ventilation. Large openings which stimulate large air movements can also provide effective cooling for people if necessary. Ceiling 'Punka' fans can also provide cooling via air movements on windless days.

Providing Efficient Lighting

Optimizing daylighting

Increasing daylighting in rooms and corridors will reduce the use of artificial lighting, but must be balanced against greater heat loss and unwanted solar gain. High-level windows possibly facing south if solar gain is required (e.g. in an east–west-facing house) can give good daylighting, as can roof lights and light tubes. Opening up windows between rooms and into corridors or halls, and using glazed doors, can provide useful 'borrowed light' without any energy penalty. Light-coloured internal painting will also improve lighting levels.

Reference: *Daylighting in Urban Areas: A Guide for Designers* (Energy Saving Trust, 2007)

Efficient lamps and luminaires

Tubular or compact fluorescent (with electronic control gear) and light-emitting diode (LED) lamps are the preferred lighting sources for all lighting in dwellings in order to minimize energy use. Light fittings or luminaires to suit these efficient lamps can be used and designed for different spaces.

Reference: *Energy Efficient Lighting: Guidance for Installers and Specifiers* ((Energy Saving Trust, 2006)

Figure 6.21 Light-emitting diode (LED) lamps are available

Source: Simon Burton

Switching

Each lighting fitting should have its own switch. Dimmer switches can only be used with tubular and four-pin florescent lamps (as well as with incandescent lamps), and will reduce energy use. Internal public spaces such as hallways should be fitted with push-button, presence-detection control and/or time-control switches. External lighting can be controlled by a time clock, presence detector or photocell controls.

Reference: *Energy Efficient Lighting: Guidance for Installers and Specifiers* ((Energy Saving Trust, 2006)

Installing Efficient Appliances

The use of electricity by appliances in low-energy homes will comprise a large proportion of household energy, and fitting low-energy A-rated appliances as part of refurbishment will reduce consumption as long as these appliances stay in place. Fitting gas hobs and cookers will generate lower carbon emissions than electric cookers, while microwaves are the most efficient cookers.

Installing Equipment to Minimize Water Use

There are a number of ways to reduce water consumption in dwellings:

- Install low-flush toilets.
- Install showers with low-flow aerated showerheads and avoid baths.
- Install spray or restricted-flow aerated taps where they are used for direct washing, rather than filling a basin.

In order to reduce mains water use, rainwater or greywater collection and storage can be installed to flush toilets and for watering the garden.

References: *Water Efficiency Retrofitting: A Best Practice Guide* (Waterwise, 2009); *Water Efficiency in New Developments: A Best Practice Guide* (Waterwise, 2010)

Reusing Existing Components and Using New Sustainable Materials

Sustainable refurbishment will reuse as much of the existing building fabric as possible, recycle material removed (e.g. as hardcore), and select new materials and components with minimal embodied energy and minimal environmental impact in manufacture. Reusing bricks and roof tiles from other demolished buildings, and use of cellulose insulation made from recycled sources are good examples. All timber should come from certified sustainable sources, with special care taken with laminated and multi-wood materials.

Comparison of the environmental credentials of different products is complex, with a variety of claims made by different manufacturers. Assessment should include energy in manufacture from raw materials to delivery, proportion of recycled material contained, environmental damage due to extraction and manufacture, longevity, maintenance, and recyclability.

Summary of the Risks and How to Avoid Them

Sustainable refurbishment can be accompanied by risks and designers need to be aware of these and ensure that risk-avoidance advice is followed. The risks associated with most elements are well known and avoidance strategies will be included in information and installation instructions. In addition, poor workmanship can always bring risks. The main areas of risk in sustainable refurbishment are as follows:

- *Interstitial condensation*. Internal insulation will reduce the temperatures of a building's structural and external elements, and this can lead to water vapour condensing on the cold surface, with consequent mould growth or deterioration of materials. This is unlikely to be a problem if there is a ventilated space between the insulation and the cold surface (e.g. in a pitched roof insulated at ceiling level) as the water can escape. Where no ventilated air gap exists, a vapour barrier should be used, placed on the warm side of the insulation. This will reduce the movement of water vapour through the insulation and lessen the interstitial vapour pressure. The important point about vapour barriers is that they must be continuous, with joints and service entries sealed and proper edge sealing so that little water vapour can pass through. Some insulation materials themselves are impervious to water; but joints and any perforations must be carefully sealed. Unventilated voids should be avoided wherever possible.

- *Increased internal condensation due to reduced ventilation.* Sealing of cracks and draught-stripping will reduce natural ventilation inside the dwelling and this can result in condensation and poor internal air quality if adequate ventilation is not provided. Natural ventilation with window ventilators, passive stack or extract ventilation can normally be adequate. Whole-house mechanical ventilation with heat recovery is becoming more common and is a requirement of the *PassivHaus* standard.

- *Cold bridging.* Where good insulation levels are applied in a house, uninsulated areas such as window frames and reveals and party walls can become cold bridges and attract condensation when internal humidity is high, which can lead to damp and mould growth. Cantilevered balconies can result in serious cold bridges and are difficult to treat. Using insulated window frames, applying some insulation to reveals, returning insulation along party walls, and insulating any mechanical fixings will overcome this problem.

- *Combustible insulation materials catching fire.* Some insulation materials are combustible and some can give off fumes, which can be dangerous both inside and outside dwellings. Combustible external cladding needs to have fire breaks at regular intervals, and internal combustible material giving off noxious gases will require normal plasterboard protection.

- *Frost damage to old materials.* Internal insulation of walls will prevent heat transfer to external walls, making them colder at certain times; it has been suggested that this could lead to more frost damage. Specialist advice is needed where damage is suspected.

- *Freezing of exposed water pipes and tanks.* Water pipes and tanks which previously benefited from adventitious heat gains from adjacent spaces will need to be insulated in order to prevent freezing if this heat supply is cut off by the installation of insulation. This is common in the case of tanks and pipes in loft spaces.

- *Overheating of electric cables.* Where insulation is likely to be placed on top of electric cables, the cables should be relocated outside the insulation in order to prevent them from becoming overheated.

- *Ventilation requirements of open-flued appliances.* If any open-flued heating appliances and those that draw combustion air from the room, such as gas heaters, coal/log-effect fires or wood stoves, exist or are to be used in the dwelling, any ventilation changes must take into account the ventilation requirements of these appliances.

- *Summer overheating.* Occasionally, sustainable refurbishment can lead to potential overheating, although this has not been recorded as a problem in the case studies presented in this book. Internal insulation can reduce the accessible thermal mass in a dwelling and thus inhibit the coolth storage ability of the house. Envelope insulation will reduce the building's night cooling and thus maintain internal temperatures during summer days. Any increased south-facing glazing could lead to excessive solar gain during a hot summer. The heat recovery part of ventilation systems should clearly not operate in summer. Relatively simple action can be taken to avoid these risks.

7 Design Tools, Testing, Monitoring and Smart Metering

Calculating energy use in existing dwellings is an essential tool when deciding which elements of low-carbon refurbishment to include and what levels are needed to achieve an overall standard. Calculation methods have to use standard occupancy and weather conditions, and actual energy use when the building is occupied is likely to vary considerably. However, calculation methods provide a basis for rational comparison between options under typical conditions and should always be used when planning a sustainable refurbishment.

Consideration should also be given to the long term since intervals between major house refurbishments tend to be long, possibly 40 years or more. Insulation of walls and floors, and frequently ceilings, is likely to last, unchanged, for many years. Therefore, a whole-life assessment of the building is always worthwhile, taking into account annual energy use, as well as maintenance and long-term replacement. Overall environmental assessment schemes may also be used to assess and demonstrate the overall sustainability of a dwelling, and to demonstrate the improvement brought about by a sustainable refurbishment project.

On-site construction needs to be monitored and, ideally, assessed in order to make sure that the design is properly implemented; the most common methods involve testing airtightness and the use of thermographic imagery.

The real test of the success of the refurbishment occurs during the occupation phase, when fuel bills and occupant satisfaction emerge. These show up any defects in the building's design or construction, as well as how the occupant is using the dwelling. Smart metering can be particularly useful to the occupant if properly used.

Energy Calculation Tools

Most countries have national energy calculation methods and these are used to comply with new-build requirements and design variations when assessing the energy usage of existing buildings. An audit of the property provides the basis of inputs to the programmes; different renovation scenarios can then be applied as individual measures and/or packages to determine the performance of the refurbished dwelling. The use of national programmes is often advisable as they are likely to be most representative of local construction and occupant use, and enable local standards (such as building regulations) and conditions (such as funding) to be met without multiple modelling.

However, where refurbishment is being designed to high insulation standards, the PassivHaus Planning Package (PHPP) may be the best tool. Its relatively uncomplicated spreadsheet tool allows designers to minimize energy use by taking into account all elements while optimizing comfort. This is obviously the tool needed to comply with if the objection is certification to *PassivHaus* standards. The PHPP includes tools for:

- calculating the U values of components with high thermal insulation;
- calculating energy balances;
- designing comfort ventilation;
- calculating the heating load;
- summer comfort calculations; and

- other tools for the reliable design of passive houses.

The 2007 PHPP is available from the PassivHaus Institute in several languages (see www. passiv.de).

In well-insulated houses, cold bridging becomes a very important issue and may be particularly difficult to deal with in renovating existing buildings. THERM is a Windows-based computer program for modelling two-dimensional heat-transfer effects in building components such as windows, walls, foundations, roofs, doors, etc. where thermal bridges are of concern. THERM is available from the Environmental Energy Technologies Division, E. O. Lawrence Berkeley National Laboratory, at the University of California.

Whole Life-Cycle Assessment

Life-cycle assessment methods abound and many are very complex, going into great detail. For use in refurbishing housing, life-cycle cost methods need to take into account capital, running and replacement costs over the lifetime of the dwelling, and life-cycle energy methods need to take into account embodied and in-use energy for the lifetime of the building. Their use may be to demonstrate that simple payback, particularly in cost terms, is not the only way to demonstrate the value of sustainable refurbishment (e.g. in relation to high insulation levels).

Overall Environmental Assessment Schemes

Since environmental assessment methods always need to evaluate, quantify and possibly rank a large number of environmental factors, there are several different schemes available, many of which are used mainly on a national basis. The following are given as examples.

LEED (Leadership in Energy and Environmental Design) for Homes is a US system that measures a home's performance based on eight categories: site selection; water efficiency; mate-rials and resources; energy and atmosphere; indoor environmental quality; location and linkages; awareness and education; and innovation. Within each of these areas, projects earn points towards certification. Each LEED-certified home undergoes on-site inspections and thorough testing to ensure proper performance. The system is applicable to single-family and multi-family homes and is intended for both owner-occupied and social or affordable housing.

In the UK, the main environmental assessment method is Ecohomes, a version of BREEAM (BRE Environmental Assessment Method) for homes. It provides a rating for new, converted or renovated homes, and covers houses, flats and apartments. Ecohomes balances environmental performance with the need for a high quality of life and a safe and healthy internal environment.

In France, the *Haute Qualité Environnementale* (HQE) is a method developed by ADEME (Agence de l'Environnement et de la Maîtrise de l'Energie), the national energy agency, and is described as an approach to managing the environmental quality of the construction and refurbishment. It is neither a regulation nor a label: it is a checklist approach that does not result in any ranking.

On-Site Monitoring and Testing

Much sustainable refurbishment work may be new to contractors; in addition, great care is needed to ensure that insulation is comprehensive and does not leave cold bridges, and that airtightness is improved upon in comparison with the original building. Inevitably, this requires high skill levels of on-site personnel, careful checking and post-construction testing. Incomplete insulation, leaving gaps, is very difficult to remedy and thus must be checked during installation before it is covered up. Thermographic imaging carried out by a professional is a good way to see how successful insulation and cold bridge avoidance has been, and possibly to determine what further work is needed.

Figure 7.1 Thermographic photography

Source: Andy Simmonds

Airtightness testing using a blower door is commonly used to assess air leakage and also tends to be a tool for progressive closing of air leakage paths if a particular target is required to be met, as with the *PassivHaus* standard.

Energy-Use Monitoring

The new generation of smart meters are helpful in providing accurate and real-time information on energy use in an understandable form to occupants. However successful the sustainable design and refurbishment has been, the actual energy use in the dwelling will still depend upon how the occupant uses the systems and appliances, and, of course, upon the weather. Energy monitoring is important for three reasons: first, for the designer to ensure that the dwelling is performing as intended, to discover any problems and to learn lessons for future work; second, so that the occupant can see that the extra sustainability works and costs are paying off in reduced fuel bills; and, third, so that the occupant can learn how to minimize their energy use by

Figure 7.2 Blower door equipment

Source: Paul Jennings

more efficient living and operation of the dwelling. Monitoring and reporting on actual energy use by reading meters should ideally be carried out for two years; during the first year the house may be drying out and the occupants will still be getting to know how to operate the heating and ventilation systems. The second year will provide a more reliable analysis. It is important to have a record of external temperatures so that variations in weather can be taken into account during the analysis, and 'degree day' adjustments made, as necessary, to standardize results.

Occupant surveys and feedback are useful for designers; but care must be taken to ensure that responses are not overshadowed by factors unrelated to sustainability issues, such as an increase in rent.

8 Adaptation to Climate Change

Some basic actions to avoid overheating are described in Chapter 6, and it makes sense to implement these as part of current sustainable refurbishment projects. This chapter looks ahead and discusses broader actions that are likely to be necessary as climate change and, particularly, associated global warming raise temperatures to new highs, making houses unbearably hot. The exact effects of climate change on different regions are uncertain, and as well as negative effects, there may also be positive ones, such as warmer winters and longer growing seasons in northern climates. For the existing housing stock, the important message is to allow and enable adaptation to take place as it becomes necessary in order to maintain comfortable living conditions. Under some scenarios, such as the disruption of the Gulf Stream, which maintains

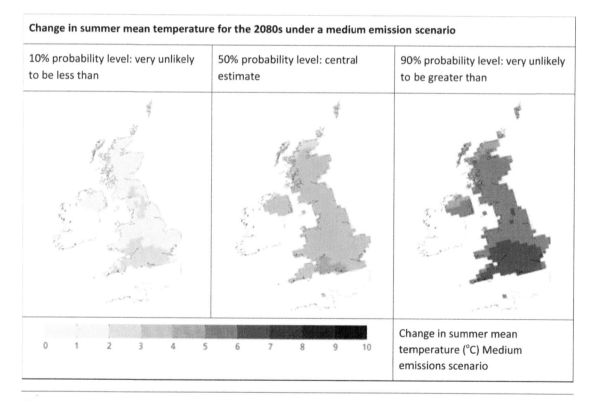

Figure 8.1 Change in summer mean temperature for the 2080s

Source: UK Climate Projections (http://ukclimateprojections.defra.gov.uk, 2009)

the UK's temperate climate, colder weather is anticipated in some regions; but a well-insulated and well-ventilated house is ideally suited to providing comfort under these conditions with minimal energy use. Similarly, some aspects of a sustainable house (e.g. insulation) can already help in combating overheating, although further actions may be necessary depending upon different weather conditions.

The Scale and Locations of the Problem

In Southern Europe, the US and similar regions, overheating of houses during summer is currently widespread and cooling using refrigeration units is the common solution for those who can afford it. This, of course, has the effect of increased electricity use and the knock-on effects of direct warming of the local environment as a result of heat dumping, increasing heat island effects in urban areas. Athens is a particularly well-documented example of this. For those unable to install air conditioning or to pay the running costs, and those concerned about further exacerbating climate change, the solution must be adaptation of the existing buildings and lifestyles of inhabitants.

Global warming seems likely to increase for the foreseeable future and this will mean that more northern regions will experience similar summer overheating, where there is no previous experience of adaptation to it. The events in France during the summer of 2008 demonstrate the serious effects of heat waves on local populations unable to adapt. Temperatures in southern regions are also forecast to rise, exacerbating the problems and requiring further adaptation.

Not a New Problem:
Learning from Hotter Regions

Communities living in much hotter climates have for many years before the invention and wider use of active mechanical cooling managed to optimize local cooling opportunities and adapt their lifestyles to excessive heat. The use of

mechanical cooling is becoming more common for affluent societies in the Middle East; but old passive-cooling techniques can demonstrate a more sustainable approach. These techniques focus on a structure that keeps out the direct sun by shading and reflective painting, on using the movements of wind and forced air, on night cooling, and on using the thermal mass of the ground to provide coolth and cool air. Lifestyle adaptation methods include resting during the hottest part of the day, using outside shaded courtyards with vegetation and water features, dressing appropriately to remain cool, and using different rooms in the dwelling according to the temperature.

Many of these passive methods are commonly used worldwide; when combined with modern materials and developments, they can certainly be used to avoid active cooling and provide a comfortable environment under a range of circumstances in existing dwellings and urban areas.

Using the Advantages of the Existing Housing Stock

The existing housing stock in most countries varies considerably depending upon age and local building design and quality. However, much of the older stock, in particular, may include features that can be used or adapted to prevent or reduce overheating. These are heavyweight construction, larger volumes and higher ceilings than modern housing, and smaller windows and sash windows, which allow high- and low-level ventilation. In other ways, older buildings may exacerbate overheating by being built with lightweight construction (e.g. 1960s housing in the UK), by having no insulation, by needing more electric lighting due to lack of daylighting, by having poor ventilation, and by having internal heat gains from inefficient water heating and electrical appliances. The best strategy is to exploit the positive features, modify the negative features, and add in new components and systems designed to reduce overheating

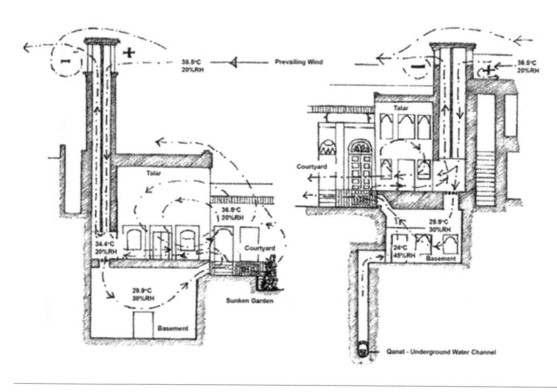

Figure 8.2 Passive cooling in Arabia

Source: Sue Roaf

Advanced Action to Avoid Overheating

Chapter 6 identifies three actions to address overheating: minimization of heat gains, the use of thermal mass and good controllable ventilation. In order to adapt to future temperature rises and to enable the dwelling to remain habitable during hot periods, all of these actions are vital and need careful and comprehensive attention. In summary, the basic requirements are that:

- all exposed east, south and west windows, as well as roof lights, have external solar shading;
- there are no large internal heat gains, such as from hot water;
- the dwelling has a lot of exposed internal thermal mass;
- the mass is insulated against the sun by external insulation;

- the ventilation is sufficient and effective, and can be used for night cooling.

In addition, further actions can be employed.

The importance of a green environment

Extensive landscaping around the dwelling will provide shading and cooler ventilation air, and enable occupants to move outside the house when it is hot inside. This applies to both private gardens and public areas. Hard paved areas are better replaced by grass or planted areas in order to prevent reflected heat and to decrease the local air temperature due to evapotranspiration. Trees, especially those that are mature, are very effective at cooling local air, and reductions in temperature of up to 3°C are claimed. Where shade and evapotranspiration are combined in groves of trees, air temperatures can be 5°C cooler than adjacent bare areas.

Landscaping can be integrated within buildings by creating green roofs and living walls. Green roofs have been demonstrated to cool buildings during summer by providing evapotranspirative cooling, reflecting sunlight and shading the building. They also have a role as a component in a sustainable drainage system, absorbing and storing rainfall, which then becomes available for evaporative cooling. The reduction in summer midday temperatures in rooms beneath green roofs, when compared with similar rooms without green roofs, has been shown to be between 5°C and 7°C. A green or living wall can also be used to reduce interior temperatures on hot summer afternoons. Fountains and similar water features also provide cooling to the local environment.

Green roofs and walls will require a design that considers structural and water penetration aspects, in particular, and will not be applicable to all dwellings; however, landscape planting and modification of surfaces near and around dwellings will usually be possible, and will always be beneficial for cooling the local environment during hot periods, with consequent cooling effects on adjacent dwellings.

Figure 8.3 A living wall

Source: Simon Burton

Ground cooling

As the ground always stays cooler than maximum air temperatures during summer, it can be a source of cooling. Although the ground, even below 1m in depth, will warm up during summer, especially if exposed to the sun, it is unlikely to go above 20°C, while the external temperature can easily reach well above 30°C. The traditional method of using this coolth in Arabic countries was to draw air from cellars when necessary for supply to living spaces. Recent experiences in France, Portugal, Germany and Greece have demonstrated good results in terms of peak temperature lopping. Air is slowly drawn by fans through a network of plastic pipes buried at least 1m below the open ground and then supplied to living spaces. The systems are best operated intermittently so that the ground has time to cool down, ready for the next demand period, and not used continuously, when the temperature of the ground may have increased considerably. It has been suggested that if there are low temperatures at night, the ground can be refreshed by running the fans at night and expelling the heat, ready for the next day.

Circulation of water in buried pipes is commonly used as a coolth source for chillers and air-conditioning systems in commercial buildings, and can also be used for dwellings. A horizontal network of pipes or a vertical borehole has been used, and these can also provide a heat source for heat pumps in winter used for space heating.

Behavioural changes

The behaviour and habits of occupants will have a significant effect on their experience of overheating. The basic issues of managing the house to keep temperatures low – such as turning off lights and unused appliances, using moveable shading, and regulating ventilation during the day and at night – are very important, as is the availability of cool and shady outdoor space for use at times of peak temperatures.

Figure 8.4 Ground cooling pipes in Portugal

Source: Simon Burton

9 Towards Zero Carbon in the Existing Housing Stock

There have been calls for all housing to become 'zero carbon' in the future – that is, responsible for no carbon emissions through the use of fossil fuels, measured over a yearly basis – with the UK government in 2010 proposing a target date of 2050. The ultimate goal of zero-carbon developments and dwellings is attractive; but even for new dwellings it is expensive and relies on extensive use of renewable energy sources. All dwellings use energy for lighting, warm-up heating and appliances; depending upon the definition of zero carbon adopted (although this can be minimized by careful design and operation), the residual energy must be provided by zero-carbon sources such as biofuel, photovoltaics, hydropower, ocean power or wind, locally or nationally. For existing housing in existing areas, zero carbon will require resources and effort beyond what is currently available. This chapter looks at how the existing housing stock can move towards the concept of zero carbon, what options are available and what needs to be developed.

Demand Reduction

Most of this book is about reducing the demand for energy use in existing housing, and this is the basic building block for the move towards zero-carbon housing. Refurbishment of dwellings according to best-practice and *PassivHaus* standards will, in the long term, be the most economic and reliable solution. Combined with the choice and use of appliances, and careful occupant control of heating, ventilation and cooling, this will minimize energy demand and reduce the energy required to be delivered from renewable energy sources.

Local and National Renewable Energy and Delivery Methods

Development and application of renewable energy sources can occur at the individual house level, local area level, and at national and international levels. Houses can be fitted with solar thermal heating, photovoltaic panels, biomass burners, ground-sourced heating and wind turbines. Local area supplies of heat and power from biomass heating, biofuelled combined heat and power (CHP), geothermal sources, wind turbine and hydroelectric sources can all be tapped into. National-level renewables supplied via the electricity grid are ocean/tidal, geothermal, wind and hydroelectric sources, solar thermo-electric stations, photovoltaic farms and biomass power stations.

National and large-scale renewables are outside the scope of this book; but building integrated renewables are very relevant, as are the issues associated with connecting to locally available supplies of renewable energy. These are discussed in more detail below.

Although it may be tempting to claim zero-carbon housing via the use of green electricity supplied via a national grid, there is widespread belief that the greening of electricity grids is a very long-term goal and that due to demand for electricity from all sectors, true zero-carbon developments must be based on local renewable supplies.

Building Integrated Renewables

Solar water heating is a tried and tested technology that can supply more than 50 per cent of

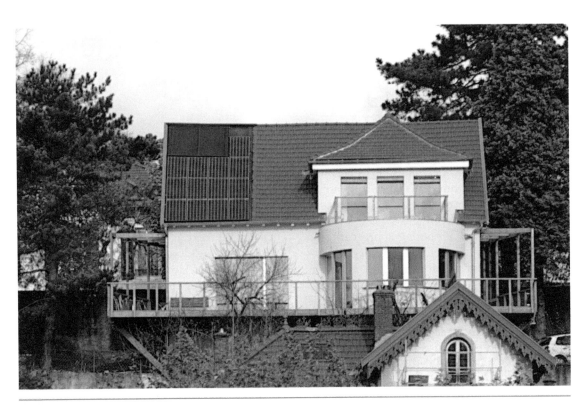

Figure 9.1 Solar water heating and photovoltaics can be combined

Source: Sophie Zele

the annual water heating demand of a dwelling, and this is an important component of a zero-carbon dwelling. Even if there is the option of heat available from a biomass boiler or biofuel district heating, a solar water heating system may be used in combination with these to supply the heating during summer when the other systems can be turned off. Evacuated tube collector systems are more effective at producing hot water with lower radiation levels, and are there-fore preferred for zero-carbon dwellings, particularly in northern regions. For grouped dwellings, such as blocks of apartments, communal solar thermal systems can be retro-fitted.

Photovoltaic (PV) arrays can be fitted to existing south-oriented roofs (as well as walls in some circumstances) or preferably integrated within replacement roofs. Electricity is generated whenever there is reasonable daylight, although direct sunlight generates the most. Panels must be free from shading as even a small amount of shading can drastically reduce the output. Large areas of PV are recommended, with electricity not used in the house being fed into the national grid. Ideally, the zero-carbon dwelling will have all south-facing roofs covered in PV and solar thermal panels. Some refurbishment projects have created south-facing roof areas in order to enable the installation of PV arrays. Arrays can equally be installed on south-facing façades, balconies and roofs of blocks of apartments, and frames can be used on flat roofs to improve the collector angle.

Figure 9.2 Large photovoltaic arrays are needed

Source: Simon Burton

Biomass burners may be appropriate for zero-carbon housing; however, a low-carbon fuel (energy is used frequently in growing and, certainly, in harvesting and delivery) should not be seen as a substitute for insulating to minimize heat demand, as the supply of biomass is always limited and it is a precious resource that has many uses in a low-carbon economy. Central heating systems may not really be necessary in dwellings insulated to *PassivHaus* standards; but biomass boilers can be an efficient way to supply space and water heating in less well-insulated dwellings. Wood-burning room stoves can be used to supply a small amount of heat to a house when needed.

Biomass heating is really only applicable to a limited number of dwellings due to fuelling issues, fuel delivery and storage, smoke emissions on start-up and ash removal. These factors make biomass burners unsuitable for sustainable housing in urban areas, small houses and apartments, and for elderly or disabled occupants.

Building-mounted wind generators have been hailed as an important renewable energy source for dwellings due to the ease of mounting them in an elevated position and the direct use of electricity by occupants. Unfortunately, their potential has not been fulfilled due to issues of low wind speeds, particularly in urban areas, turbulence, poor technology, the need for maintenance, and noise. In addition, house-mounted generators are inevitably of a small size with a small maximum output compared to stand-alone generators; even under ideal conditions, annual

Figure 9.3 A wood-burning stove

Source: Simon Burton

contributions have proved to be small.

Heat pumps can have a role in the zero-carbon dwelling where heat demands are small and where electricity is ideally available from renewable resources. Ground-sourced heat pumps have a better output and efficiency than air sourced, and they are normally considered to be a renewable energy option. In hot climates where active cooling is often used, ground cooling can be an effective sustainable alternative, as described in Chapter 8.

Local Renewables and Delivery Networks

Where a group of dwellings is being sustainably renovated and is moving towards zero carbon, the opportunity may arise to provide renewable energy sources communally. Blocks of flats can have communal solar water systems, PV arrays mounted on roofs, wind turbines benefiting from a high location, or even larger turbines mounted in car parks and other adjacent areas, as well as communal biomass heating systems. The economics of communal systems can be better than smaller individual systems; usage of available energy may be more efficient and maintenance can be more easily organized.

At a wider scale, district heating and private electricity networks enable the direct use of energy from renewable resources adjacent to housing. These sources include biomass-fired power stations, large wind turbines or farms, hydroelectric generators, or even PV arrays and solar electric farms or ocean/tidal devices. Where

Figure 9.4 District heating networks are a way of providing joint renewable energy supplies

Source: Ramboll UK

such networks are available, connection is likely to be the best way of supplying energy to a sustainable refurbishment project, although private-wire electricity networks would appear to have few advantages over the use of national grids.

The development of new district heating networks based on renewable heat sources is being promoted by some governments and other organizations, and may be viable and beneficial in built-up areas. Traditionally, district heating networks, at both large and small (block) scales, have been seen as essential for the wide-scale use of renewables, CHP and 'waste' heat, as we move towards a society based on more sustainable energy use. However, there are several factors that militate against the widespread application

of district heating for areas of existing housing which is to be renovated to zero carbon. The fact that the demand for both space and water heating after sustainable refurbishment should be low and intermittent, the cost and disruption of installing the network, the heat losses from the pipes in the network, the need for supply to satisfy different demand profiles, and maintenance aspects all mitigate against district heating compared with dwelling-based renewable heating or the use of fossil-fuel heating or electricity from a renewable source.

Private-wire electricity supply networks are sometimes proposed as part of renewable energy or CHP developments; but since virtually all dwellings are already (and will always need to be) connected to the national grid, the development of a parallel system does not offer benefits in terms of the use of renewable energy, which can just as easily be fed into the national grid with appropriate metering and allocation agreements.

Achieving a Sustainable Future

Zero-carbon homes, both new and existing, are technically feasible if one takes the broader definition of associated renewable energy sources. For the individual house owner, architect or housing planner, minimizing the energy footprint of a dwelling by reducing demand and maximizing the use of on-site renewable energy sources is the goal, and this can produce very low-carbon homes. To achieve this on a large scale, all participants need to understand and think sustainably; the owner needs to see the benefits and have access to adequate finance; the architect needs to design sustainably; the local planner needs to enable (and encourage) sustainability; the finance organization needs to fund sustainability; the builder needs to construct sustainably; suppliers of building materials need to recognize sustainability in the supply chain and stock sustainable products; and, finally, the occupier needs to run the house sustainably. In other words, everyone has to take on sustainability as a component of their daily activities.

Urgent action is needed and this has started in many areas. Best practice examples exist, design and building expertise is growing, building physics is well advanced, suitable products and materials are available, national and local governments are providing support, and the general public is perhaps awakening to the need for sustainable housing and living. Many more years of concerted effort are still required.

Index

Printed and bound by CPI Group (UK) Ltd, Croydon, CR0 4YY

23/10/2024

01777684-0001